CLASSICS OF SEA POWER

SERIES EDITORS' INTRODUCTION

The Classics of Sea Power series makes readily available, in uniform, authoritative editions, the central concepts of the naval profession. These major book-length works in the words of the masters have been chosen for their eloquence and timelessness, and express the important themes of strategy, operations, tactics, and theory.

With Alfred Thayer Mahan, the seventh author in the series, we come to the man whose very name embodies the notion of sea power. Most naval officers, however, and many historians and statesmen are acquainted with Mahan only through the interpretations of secondary sources. So that Mahan be seen neither as god on a pedestal, too remote, deep, and profound to be approached directly, nor as a mere commentator of a day now 100 years past relevance, we offer every reader the chance to drink the real man from his own cup. In particular the reader may find for himself here the manner and extent of Mahan's espousal of sea power, both as it offered greatness and threatened vulnerability to a state. One may also discover the firm, forthright, yet nuanced and correlated, way he saw the battle fleet, command of the sea, and the safe movement of shipping as a triad of which maritime achievement is based.

Mahan should neither be idolized nor demythologized, but the subject of study, lest it be his ghost and not the writer himself that influences the modern American strategist.

SERIES EDITORS

John B. Hattnedorf
Naval War College
Newport, Rhode Island

Wayne P. Hughes, Jr.
Naval Postgraduate School
Monterey, California

Mahan on Naval Strategy

Rear Admiral Alfred Thayer Mahan

REAR ADMIRAL
ALFRED THAYER MAHAN,
U.S. NAVY

Mahan on Naval Strategy

Selections from the Writings of Rear Admiral Alfred Thayer Mahan

With an Introduction
by John B. Hattendorf, Editor

NAVAL INSTITUTE PRESS Annapolis, Maryland

This book has been brought to publication with the generous assistance of Marguerite and Gerry Lenfest.

Naval Institute Press
291 Wood Road
Annapolis, MD 21402

These writings of Alfred Thayer Mahan have been reprinted from original sources indicated on the opening pages of each chapter.

First Naval Institute Press paperback edition published in 2015.
ISBN: 978-1-59114-559-2 (paperback)
ISBN: 978-1-61251-819-0 (eBook)

The Library of Congress has cataloged the hardcover edition as follows:
Mahan, A. T. (Alfred Thayer), 1840–1914.
Mahan on naval strategy : selections from the writings of Rear Admiral Alfred Thayer Mahan / Alfred Thayer Mahan : with an introduction by John B. Hattendorf, editor.
p. cm.—(Classics of sea power)
Includes bibliographical references and index.
ISBN 1-55750-556-X (acid-free paper)
1. Naval strategy. I. Hattendorf, John B. II. Title. III. Series
V163.M19 1991
359'.03—dc20

90-25401

♾ Print editions meet the requirements of ANSI/NISO z39.48-1992 (Permanence of Paper).
Printed in the United States of America.

26 25 24 23 22 21 13 12 11 10 9 8

Frontispiece courtesy of the U.S. Naval Institute Photo Archive.

CONTENTS

Editor's Introduction ix
Introductory 1
Discussion of the Elements of Sea Power 27
Foundations and Principles 97
Strategic Lines 142
Distant Operations and Maritime Expeditions 177
Operations of War 219
Considerations Governing the Disposition of Navies 281
The Persian Gulf and International Relations 319
The Naval War College 343
Britain & the German Navy 359
The Panama Canal and the Distribution of the Fleet 368
Index 383

EDITOR'S INTRODUCTION

ALFRED THAYER MAHAN made two great contributions to the development of naval thought. First, he linked maritime and naval activities to wider national and international issues. After his series of sea power books, students of naval affairs were no longer satisfied with mere descriptions of battles at sea, as they had previously been. They looked for wider implications and interrelationships. Secondly, he laid out a series of principles for professional naval officers to use in the formulation of naval strategy. His adaptation and refinement of Jomini's approach to military science marked an intellectual revolution for navies; after reading Mahan's works, naval officers had found a tool with which they could develop strategic naval doctrine.

Mahan thought that his two contributions were of equal importance, each interdependent on the other. Despite the author's own view, scholars have paid more attention to Mahan's first contribution than to his second. Mahan's ideas on sea power have been more interesting to historians as they debated explanations of national and international history. At the same time, naval officers have tended to be more attracted to those concepts in Mahan's writings that can be readily applied to future situations. On the one hand, these contributions seem to fall into the province of historical scholarship; on the other, into the realm of political science.

Different as these two are, they are, however, closely tied to one another. The linking feature is Mahan's approach to the study of history. As Mahan wrote:

> Formulated principles, however excellent, are by themselves too abstract to sustain convinced allegiance; the reasons for them, as manifested in concrete cases, are an imperative part of the process through which they really enter the mind and possess the will. On this account the study of military history lies at the foundation of all sound military conclusions and practice.[1]

By making his dual contribution to naval thought, Mahan's work became part of a growing evolution of ideas, which included those of both his precursors and successors. In the application of historical study to the development of naval doctrine, he was clearly following the lines that Sir John Knox Laughton had laid out in England in the 1870s. Rear Admiral Stephen B. Luce, the founder of the Naval War College at Newport, Rhode Island, transmitted to Laughton's ideas to Mahan,[2] along with Luce's own wide approach to the study of the highest branches of naval thought. Luce's ideas led directly to Mahan's initial series of lectures at the Naval War College. In book from, these lectures eventually appeared as *The Influence of Sea Power Upon History 1660—1783* (1890) and *Naval Strategy Compared and Contrasted with the Principles and Practice of Military Operations on Land* (1911).

There were a number of other writers whose work complemented and expanded Mahan's. First, there was Vice Admiral

1. "The Naval War College," see, p. 000.

2. On the connection between Luce and Laughton, see John D. Hayes and John B. Hattendorf, eds., *The Writings of Stephen B. Luce* (Newport, 1975), pp. 71, 84, 91, 99, 178. For further information on Laughton, see D. M. Schurman, *The Education of a Navy: The Development of British Naval Strategic Thought 1867–1914* (Chicago, 1965), chapter 5, pp. 83–109.

Philip Colomb. Like Luce and Mahan, he had also been influenced by Laughton. Colomb's book, *Naval Warfare: Its Ruling Principles and Practice Historically Treated*[3] appeared simultaneously with Mahan's first sea power book in 1890, but concentrated on identifying various types of naval operations and their uses.

Among others who succeeded Mahan, Sir Julian Corbett had the advantage of the pioneer work of Laughton as well as that of both Mahan and Colomb. Additionally, he could give a deeper intellectual dimension to naval affairs by applying Clausewitz's philosophical approach to understanding the nature of warfare.[4] Thus, Corbett went beyond Mahan, subsuming his ideas and placing them in a wider context.

From the work of Mahan, Colomb, and Corbett onward, naval theorists have continued to develop and refine naval theory.[5] In the history of this development, Mahan stands out as the writer whose work was the first to be widely read and appreciated. Through the many editions and translations of his works, and through the repetitive publication of his shorter pieces, first in newspapers, then adapted to journal articles and finally appearing as chapters in books, Mahan's ideas gained a currency that few, if any, other authors in naval affairs have achieved.[6]

3. See the new edition of Colomb's *Naval Warfare* in this series, with an introduction by Barry M. Gough (Annapolis, 1990).

4. See the new edition of Corbett's *Some Principles of Maritime Strategy* in this series, annotated with an introduction by Eric J. Grove (Annapolis, 1988).

5. For an outline of these developments, see John B. Hattendorf, "Recent Thinking on the Development of Naval Theory," in Hattendorf and Robert S. Jordan, eds., *Maritime Strategy and the Balance of Power: Britain and American in the 20th Century.* (London and New York, 1989), pp. 136–61.

6. For the details of this pattern of publication, see John B. Hattendorf and Lynn C. Hattendorf, *A Bibliography of the Works of Alfred Thayer Mahan.* (Newport, 1986; reprinted 1990).

In the history of naval thought, Alfred Thayer clearly stands out as a great figure, yet a precise understanding of his ideas has been difficult to grasp. Much of this has been due to his large output: 20 books, 22 contributions to other books, 5 volumes (including this one) of his work collected and edited by others, 161 journal articles, 109 known newspaper articles, 27 translated articles, and 13 pamphlets. To this bulk one must add Mahan's own reluctance to summarize his ideas in any abstract form. Most of his works appear to be detailed narrative descriptions, which resist a rapid assimilation of ideas without a close reading. Mahan took this approach for a very definite purpose. He firmly believed that detailed historical narrative and the statement of principles were reciprocal approaches:

> Each is a partial educator; combined, you have in time a perfect instructor. Of the two, History by itself is better than formulated principles by themselves; for in this connection, History, being the narrative of actions, takes the role which we commonly call practical. It is the story of practical experience. But we all, I trust, have advanced beyond the habit of thought which rates the rule of thumb, mere practice, mere personal experience, above practice illuminated by principles, and reinforced by the knowledge, developed by many men in many quarters. Master your principles, and then ram them home with the illustrations which History furnishes.[7]

The purpose of this volume is to select some key pieces that lay out Mahan's basic ideas, while also preserving a modicum of illustration, as Mahan would have had it.

BIOGRAPHICAL SKETCH

Several writers have shown in biographical studies of Mahan[8] that he had a typical career for a naval officer, until shortly

7. *Naval Strategy* (1911), p. 17.
8. The most recent, and one that summarizes the most useful literature,

before his fiftieth birthday, when he published his first sea power book. He was born on 27 September 1840 at the U.S. Military Academy, West Point, New York, where his father, Dennis Hart Mahan, had become well-known as professor of civil and military engineering and dean of the faculty. Over the years, the elder Mahan had become famous for his own writings and teachings based on the ideas of Jomini. The son seemed little affected by this intellectual background and, perhaps, closed it out of his mind following his father's suicide in 1871, when West Point did not retain the elder Mahan on the faculty after he had reached his late sixties.

The young Mahan went on to the Naval Academy in 1856, where he graduated with the class of 1859. As a lieutenant during the Civil War, he served in three ships on various blockade duties, and also at the U.S. Naval Academy in 1862–63, where he served under Lieutenant Commander Stephen B. Luce, then commanding the midshipman training ship USS *Macedonian.* After the war, he served in the USS *Iroquois* on the Asiatic Squadron, then commanded the USS *Wasp* and later, the USS *Wachusett,* with interim spells of duty at the navy yards in Boston and New York. During those years, the only thing that was slightly unusual was that he had published two pieces. One was an article, "Naval Education for Officers and Men," which had won him second honorable mention in the Naval Institute Prize Essay Contest for 1879. The other was a book, *Gulf and Inland Waters,* part of a series narrating the naval history of the American Civil War. There were many other officers who had written and published more extensively than Mahan, and are far less well

is Robert Seager II, "Alfred Thayer Mahan," in James C. Bradford, *Admirals of the New Steel Navy: Makers of American Naval Tradition 1880–1930* (Annapolis, 1990), pp. 24–72, and Philip A. Crowl, "Alfred Thayer Mahan: The Naval Historian" in Peter Paret, ed., *Makers of Modern Strategy from Machiavelli to the Nuclear Age* (Princeton, 1986), pp. 444–77.

known today. Although interesting and useful works, neither piece showed any clear hint of the work to come. Perhaps the most decisive factor in his career to this point was the fortunate, short assignment to Stephen B. Luce's ship in 1863. For it was Luce who would call Mahan to join him at the newly established Naval War College in 1885 to teach naval history, and who at that point became the guiding force creating for Mahan, through the Naval War College, a newly found focus and intellectual approach for the remainder of Mahan's life. As Mahan explained in 1892,

> The cordial reception . . . [of my work] has been . . . not only most gratifying, but wholly unexpected. Its chief significance is, however, not personal. The somewhat surprised satisfaction testified is virtually an admission that, in the race for material and mechanical development, sea officers as a class have allowed their attention to be unduly diverted from the systematic study of the Conduct of War, which is their peculiar and main concern. For, if the commendation bestowed be at all deserved, it is to be ascribed simply to the fact that the author has been led to give to the most important part of the profession an attention which it is in the power of any other officer to bestow, but which too few actually do.
>
> That the author has done so is due, wholly and exclusively, to the Naval War College, which was instituted to promote such studies.[9]

Promoted to the rank of captain in 1885, he came to the Naval War College in that year to begin serious work on his first sea power book, basing himself at his home in New York City. The following year, he succeeded Luce as the president of the Naval War College, serving in that capacity until 1889. In that year, the Navy Department assigned him to be the president of a commission to select a site for a navy yard on

9. *The Influence of Sea Power Upon the French Revolution and Empire* (Boston, 1892). Foreword.

the northwest Pacific Coast. After selecting the site, which became the Puget Sound Navy Yard, he served on special duty in the Bureau of Navigation at Washington, and then the Navy Department ordered him to a second period as president of the Naval War College in 1892–1893. Following this period, in which he reestablished the Naval War College, after an attempt to merge it with technical training at the Naval Torpedo School, Mahan went on to command the USS *Chicago* in 1893–1895. During a cruise in European waters while on this duty, he and his works received widespread attention. The press followed his activities closely, the universities of Oxford and Cambridge awarded him honorary degrees, and both Queen Victoria and Kaiser Wilhelm II received him. Following this last sea assignment and after forty years of service, Mahan retired from active duty in the navy to devote himself fully to his writing career. Despite his retirement, he retained his naval connections, serving as a member of the Naval War Board during the war with Spain in 1898, and on special duty with the Naval War College from 1908 until 1912. As part of recognition given to officers who had served in the Civil War, the U.S. Navy promoted Mahan to rear admiral on the retired list, with date of rank from 29 June 1906.

He died in Washington, D.C., on 1 December 1914, only a few months after the First World War had begun.

MAHAN'S INTELLECTUAL WORLD

Mahan was very much a child of his times. His writings are not among those that transcend their own era; they are very much a product of them. To understand them fully one needs to have some empathy for the period of the late nineteenth century. Even to his contemporaries, his writings were not entirely attractive. One newspaper report succinctly summarized the views of the *Manchester Guardian:*

> Captain A. T. Mahan gets an odd mixture of praise and blame from one of his English critics—the reviewer of the "Manchester Guardian" to wit. This worthy speaks of the historian's "wonderful power of exposition," but declares that he has "no skill in story telling, no power of color or of humor, no liveliness, none of that delight in detail which makes a memoir great and damns a history." As for his style it is less attractive says the commentator, than the style of any man of similar eminence. It is cold, it is heavy, it is unrhythmical; it is without any quality of beauty. But as a historian he compels admiration—he has such a grasp upon his subject; his cold, clumsy, telling phrases go home deeply. His "nuts of knowledge" are heavy round-shot.[10]

Mahan's work had a direct relevance to his own time, a time of dramatic change for the world's navies. For much of the nineteenth century, Britain's navy was the preeminent force afloat. With no serious rivals after 1815, we have come to think of the period as that of the *Pax Britannica,* although the term itself did not come into use until after 1886.[11] By that date, it was already about to come a thing of the past. The *Pax Britannica* was never something forced upon the world by the operations of a large navy. It was a situation in which other nations accepted British ascendancy in the areas of finance, industry, commerce, and shipping. No other nation had developed the industrial and economic capacity to challenge Britain, and thus, while the Royal Navy discouraged competition, other nations were willing to accept her ideas of free trade, peace, and prosperity. This situation could only exist while others did not challenge it. The age of the *Pax Britannica* was over when others began to industrialize, pro-

10. From an unidentified, contemporary newspaper clipping pasted into J. B. Hattendorf's copy of *The Interest of America in Sea Power, Past and Present* (Boston, 1898). Below this clipping, and apparently pasted in after it, is another dated 11th November 1900.

11. Barry M. Gough , "*Pax Britannica:* Peace, Force and World Power," *The Round Table* (1990), vol. 314, p. 168.

duce steam machinery, lay iron railroads, and construct iron ships from the products of their own factories.[12] That revolution began to occur in the 1880s, just as other European nations began to become interested in the competition for empire. At the beginning of the century, Britain had been arbiter of international security, but by the end of the century she replaced her global preponderance with a strategy of local concentration. While this fundamental change was occurring in the structure of international relations, another dramatic change, adding a new dimension to naval competition, affected the technological basis of navies: the change from wood to iron, sail to steam, and round shot to shells. This took place simultaneously, although it stretched over a longer period. It brought with it not only a new technological emphasis, but a new approach to naval tactics and strategy. Without bases for refueling, a navy's radius of operations was restricted, but it had greater maneuverability within an area. The new technology enhanced the capabilities for distant nations to develop local naval superiority, thus challenging the preponderance of a single, global power. With the arrival of these industrial capabilities, both the United States and Japan built navies that took advantage of the situation.

This twofold change struck directly at the traditional thinking about navies. Previously, naval leaders developed their strategy in terms of common sense and practical experience. No serious professional literature existed. There had been no need to intellectualize the subject. The changed nature of international politics and naval technology brought all into question and seemed to sweep all previous thinking away. Mahan, and the others who may be counted among the naval leaders of this movement—Laughton, Luce, Colomb, and Corbett—all looked for a means to find some enduring and

12. Gerald S. Graham, *The Politics of Naval Supremacy: Studies in British Maritime Ascendancy* (Cambridge, 1965), pp. 120–21.

steady guide amidst the constant flux that surrounded them. With the others, Mahan looked to Britain's experience as the preeminent naval power and sought to find guidance from it. Britain's use of naval power during the period between 1660 and 1815 caused her to emerge successfully from a global struggle for empire. This historical example might not have had universal validity, but it appeared clearly relevant to a new age of imperial rivalry in the 1880s and 1890s.

A TOPICAL CATALOG OF QUOTATIONS ON NAVAL STRATEGY

Mahan's thoughts touched, time and again, on several topics as he dealt with specific events in history and in commenting on contemporary strategy in the period of his writing between 1885 and 1914. The following catalog of key quotations from these writings can serve as a summary of his ideas, complementing the longer exposition of his thoughts presented in the selections reprinted in this volume:[13]

Coast Defence

"Sea ports should defend themselves; the sphere of the fleet is on the open sea, its object offense rather than defence, its objective the enemy's shipping whenever it can be found."[14]

"In naval warfare, coast defence is the defensive factor, the navy the offensive. Coast defence, when adequate, assures

13. See the general index to this volume for cross-refeiences in the text. The following quotations have been selected from the compilation by Dr. Jean Ware Nelson in "The Sea Power Doctrine of Admiral Alfred Thayer Mahan," a staff working paper prepared for the Naval Warfare Research Center, Stanford Research Institute, in December 1960 at the request of the Strategic Plans Division, Office of the Chief of Naval Operations.

14. *The Influence of Sea Power Upon History, 1660–1783* (1890; quotation from 1957 ed.), p. 404.

the naval commander-in-chief that his bases of operations—the dockyards and coal depots—is secure. It also relieves him and his government, by the protection afforded to the chief commercial centers, from the necessity of considering them, and so leaves the offensive arm perfectly free."[15]

COMMUNICATIONS

"Communications dominate war; broadly considered, they are the most important single element in strategy, political or military."[16]

"In this respect, the navy is essentially a light corps; it keeps open the communications between its own ports, it obstructs those of the enemy; but it sweeps the sea for the service of the land, it controls the desert that man live and thrive on the habitable globe."[17]

CONCENTRATION

"Like the Land, the sea, as a military field, has its important centers, and it is not controlled by spreading your force, whatever its composition, evenly over an entire field of operations, like butter over bread, but by occupying the centers with aggregated forces—fleets or armies—ready to act in masses, in various directions from the centers. This commonplace of warfare is its first principle. It is called concentration because the forces are not spread out, but drawn together at the centers which for the moment are important."[18]

15. *The Interest of America in Sea Power, Past and Present* (1897), p. 194.

16. *The Problem of Asia and Its Effect Upon International Policies* (1900), p. 125.

17. *The Influence of Sea Power Upon History, 1660–1783* (1957 ed.), p. 290.

18. *Lessons of the War With Spain* (1899), p. 258.

CONTROL

"Control of a maritime region is insured primarily by a navy; secondarily, by positions, suitably chosen and spaced one from the other, upon which as bases the navy rests, and from which it can exert its strength."[19]

COMMAND OF THE SEA

"These national and international functions can be discharged, certainly, only by command of the sea. The Pacific, the Atlantic, and the Caribbean, with the great controlling stations, Porto Rico, Guantanamo, The Canal Zone, and Hawaii, depend upon this command, the exponent of which is the navy, and in which ships and stations are interdependent factors. To place the conclusion concretely, and succinctly, the question of command of the sea is one of annual increase of the navy. The question is not 'naval,' in the restricted sense of the word. It is one of national policy, national security, and national obligation."[20]

DEFENSIVE VERSUS OFFENSIVE

"A sound defensive scheme, sustaining the bases of the national force, is the foundation upon which war rests; but who lays a foundation without intending a superstructure? The offensive element in warfare is the superstructure, the end aim for which the defensive exists, and apart from which it is to all purposes of war worse than useless. When war has been accepted as necessary, success means nothing short of victory; and victory must sought by offensive measures, and by them only can be insured."[21]

19. *The Interest of America in Sea Power, Past and Present*, p. 102.

20. "The Importance of Command of the Sea," *Scientific American*, vol. 105, 9 December 1911, p. 512.

21. *Retrospect and Prospect. Studies in International Relations, Naval and Political* (1902), p. 152.

"In war, the defensive exists mainly that the offensive may act more freely. In sea warfare, the offensive is assigned to the navy; and if the latter assumed to itself the defensive [guarding bases], it simply locks up a part of its trained men in garrisons, which could be filled as well by forces that have not their peculiar skill."[22]

"Even though the leading object of the war be defense, defense is best made by offensive action."[23]

"Not speed, but power of offensive action, is the dominant factor in war. The decisive preponderant element of great land forces has ever been the infantry, which, it is needless to say, is also the slowest."[24]

Deterrence

"It is one thing to be strong enough to make an adversary wary of attacking; it is quite another to be able to beat him if a contest arises."[25]

"Indeed, force is never more operative than when it is known to exist but is not brandished."[26]

Fleet

"A country can, or will, pay only so much for its war fleet. That amount of money means so much aggregate tonnage. How shall that tonnage be allotted? And, especially, how shall the total tonnage invested in armored ships be divided? Will you have a very few big ships, or more numerous medium ships?

Where will you strike your mean between number and in-

22. *Naval Strategy*, p. 150

23. Ibid., p. 205.

24. *Lessons of the War with Spain*, p. 82.

25. *The Interest of America in International Conditions* (1910), p. 160–61.

26. *Armaments and Arbitration, or the Place of Force in the International Relations of States* (1912), p. 105.

dividual size? You cannot have both, unless your purse is unlimited."[27]

"Among the naval entities, fleets are at once the most powerful and the least mobile; yet they are the only really determining elements in naval war."[28]

FLEET-IN-BEING

"It is indeed as a threat to communications that the 'fleet in being' is chiefly formidable."[29]

"The fleet-in-being concept exalts the defensive and is erroneous."[30]

"The fleet-in-being has a moral as well as physical effect."[31]

FORCE

"The great end of a war fleet . . . is not to chase, nor to fly, but to control the seas . . . Not speed, but power of offensive action, is the dominant factor in war . . . Force does not exist for mobility, but mobility for force. It is of no use to get there first unless, when the enemy in turn arrives, you have also the most men, the greater force."[32]

"Do not lose sight of the fact that all organized force is in degree war, and that upon organized force the world so far has progressed and still progresses. Upon organized force depends the extended shield, under which the movements of peace advance in quietness; and of organized force war is simply the last expression."[33]

27. *The Lessons of the War With Spain* (Boston, 1899), p. 37.

28. Ibid., p. 262.

29. *The Interest of America in Sea Power, Present and Future*, p. 77.

30. *The Major Operations of the Navies in the War of American Independence* (1913), p. 174.

31. *The Life of Nelson: The Embodiment of Sea Power of Great Britain* (1897), vol. 1, p. 136.

32. *Lessons of the War with Spain*, p. 83.

33. *Some Neglected Aspects of War* (1907), p. 89.

Government

"The government by its policy can favor the natural growth of a people's industries and its tendencies to seek adventure and gain by way of the sea; or it can try to develop such industries and such sea-going bent, when they do not naturally exist; or, on the other hand, the government may by mistaken action check and fetter the progress which the people left to themselves would make. In any one of these ways the influence of the government will be felt, making or marring the sea power of the country in the matter of peaceful commerce; upon which alone, it cannot be too often insisted, a thoroughly strong navy can be based."[34]

History

"There is such a thing as becoming imbued with the spirit of a great teacher, as well as acquainted with his maxims. There must indeed be in the pupil something akin to the nature of the master thus to catch the inspiration—an aptitude to learn; but the aptitude, except in the rare cases of great original genius, must be brought into contact with the living fire that it may be itself kindled."[35]

Mobility

"Naval strength involves, unquestionably, the possession of strategic points, but is greatest constituent is the mobile navy."[36]

National Policy

"Justly appreciated, military affairs are one side of the politics of a nation, and therefore concern each individual who has an interest in the government of the state. They from part of

34. *The Influence of Sea Power Upon History, 1660–1783*, p. 82.
35. *Naval Strategy*, p. 298.
36. Ibid., p. 127.

a closely related whole; and, putting aside the purely professional details, which relate mostly to the actual clash of arms—the province of tactics—military preparations should be determined chiefly by those broad political considerations which affect the relations of states one to another, or of the several parts of the same state to the common defense."[37]

"The office of statesman is to determine, and to indicate to the military authorities, the national interests most vital to be defended, as well as the objects or conquest or destruction most injurious to the enemy, in view of the political exigencies which the military power only subserves. The methods by which the military force will proceed to the ends thus indicated to it—the numbers, character, equipment of forces to be employed, and their management in campaign—are technical matters, to be referred to the military or naval expert by the statesmen. If the latter undertakes to dictate in these, he goes beyond his last and commonly incurs misfortune."[38]

"The sphere of the navy is international solely. It is this which allies it so closely to that of the statesman. Aim to be yourselves statesmen as well as seamen."[39]

Naval Strategy

"Naval strategy has for its end to found, support, and increase, as well in peace as in war, the sea power of a country."[40]

Objectives and the objective in warfare

"This of course leads us straight back to the fundamental principles of all naval war, namely, that defence is insured

37. *Naval Administration and Warfare, Some General Principles with Other Essays*. (1908), p. 137.

38. *The Influence of Sea Power Upon the French Revolution and Empire, 1793–1812* (1892), vol. 2, pp. 391–92.

39. *Naval Strategy*, p. 21.

40. *The Influence of Sea Power on History, 1660–1783*, p. 89.

only by offense, and that the one decisive objective of the offensive is the enemy's organized battle-fleet."[41]

"The harassment and distress caused to a country by serious interference with its commerce will be conceded by all. It is doubtless a most important secondary operation of naval war, and is not likely to be abandoned till war itself shall cease; but regarded as a primary and fundamental measure, sufficient in itself to crush an enemy, it is probably a delusion and a most dangerous delusion, when presented in the fascinating garb of cheapness to the representatives of a people."[42]

"Ships and cargoes in transit upon the sea are private property in only one point of view, and that the narrowest. Internationally considered, they are national wealth engaged in reproducing and multiplying itself, to the intensification of the national power, and that by the most effective process; for it relieves the nation from feeding itself and makes the whole outer world contribute to its support. It is therefore a most proper object of attack; more humane, and more conducive to the objects of war, than the slaughter of men."[43]

"The true standard of civilized warfare is the least injury consistent with the end in view; but the end should not be lost to sight in glittering generalities."[44]

Position

"It is in the utilization of position by mobile force that war is determined, just as the effect of a chessman depends upon both its individual value *and* its relative position. While, therefore, in the combination of the two factors, force and position, force is intrinsically the more valuable, it is always

41. *Retrospect and Prospect,* p. 163.
42. *The Influence of Sea Power Upon History, 1660–1783* (1957 ed.), p. 481.
43. *Retrospect and Prospect,* p. 144.
44. *Naval Administration and Warfare,* p. 105.

possible that great advantage of position may outweigh small advantage of force, as 1 + 5 is greater than 2 + 3."[45]

"The sea itself becomes a link, a bridge, a highway, a central position, to the navy able to occupy it in adequate force. It confers interior lines, central position, and communications militarily assured; but to hold it requires the possession of overseas bases, fortresses, such as those of which we have been speaking."[46]

"In naval war the fleet itself is the key position of the whole."[47]

Preparedness

"Naval strategy is as necessary in peace as in war."[48]

"Every danger of a military character to which the United States is exposed can be best met outside her own territory—at sea. Preparedness for naval war—preparedness against naval attack and for naval offense—is preparedness for anything that is likely to occur."[49]

Retaliation

"It is a grievous fault of all retaliation, especially in the heat of war, that it rarely stays its hand at an equal measure, but almost invariably proceeds to an excess which provides the other party to seek to even the scale. The process tends to be unending. . . ."[50]

"If retaliation upon any but the immediate culprit is ever permissible, which in national matters will scarcely be contested, it is logically just that it should fall first of all upon the

45. *The Interest of America in Sea Power, Present and Future*, p. 291.
46. *Naval Strategy*, p. 99.
47. *Naval Strategy*, p. 191.
48. *Naval Administration and Warfare*, p. 172
49. *The Interest of America in Sea Power, Present and Future*, p. 214.
50. *Sea Power in Its Relation to the War of 1812* (1905), vol. 2, p. 335.

capital, where the interests and honor of the nation are centered. There, if anywhere, the responsibility for the war and all its incidents is concentrated in the representatives of the nation, executive and legislative, and in the public offices from which all overt acts are presumed to emanate."[51]

RISK

"Something must be left to chance; nothing is sure in a sea flight beyond all others."[52]

". . . for only men of the temper of Farragut or Grant—men with a natural genius for war or enlightened by their knowledge of the past—can fully commit themselves to the hazard of a great adventure—can fully realize that a course of timid precaution may entail the greatest of risks."[53]

SEA POWER

"Is it meant, it may be asked, to attribute to sea power alone the greatness or wealth of any State? Certainly not. The due use and control of the sea is but one link in the chain of exchange by which wealth accumulates; but it is the central link, which lays under contribution other nations for the benefit of the one holding it, and which, history seems to assert, most surely of all gathers to itself riches."[54]

"If navies, as all agree, exist for the protection of commerce, it inevitably follows that in war they must aim at depriving their enemy of that great resource; nor is it easy to conceive what broad military use they can subserve that at all compares with the protection and destruction of trade."[55]

51. Ibid. p. 336.

52. *The Life of Nelson,* vol. 2, p. 344

53. *Admiral Farragut* (1892), p. 223.

54. *The Influence of Sea Power Upon History, 1660–1783* (1957 ed.), p. 200.

55. *The Interest of America in Sea Power, Present and Future* (1897), p. 128.

"I am not particularly interested here to define the relations of commerce to a navy. It seems reasonable to say that, where merchant shipping exists, it tends logically to develop the form of protection which is called naval; but it has become perfectly evident, by concrete examples, that a navy may be necessary where there is no shipping. Russia and the United States today are such instances in point. More and more it becomes clear, that the function of navies is distinctly military and international, whatever their historical origin in particular cases."[56]

"The noiseless, steady, exhausting pressure with which sea power acts, cutting off the resources of the enemy while maintaining its own, supporting war in scenes where it does not appear itself, or appears only in the background, and striking open blows at rare intervals, though lost to most, is emphasized to the careful reader by the events of this war [The War of the Spanish Succession, 1702–1714] and of the half century that followed."[57]

"The world has never seen a more impressive demonstration of the influence of sea power upon its history. Those far distant, storm-beaten ships, upon which the Grand Army never looked, stood between it and the dominion of the world."[58]

"Sea power, however, is but the handmaiden of expansion, its begetter and preserver; it is not itself expansion, nor did the advocates of the latter foresee room for advance beyond the Pacific."[59]

56. *Naval Strategy*, p. 446–47.

57. *The Influence of Sea Power Upon History, 1660–1783* (1957 ed.), p. 186.

58. *The Influence of Sea Power Upon the French Revolution and Empire, 1793–1812* (1892), vol. 2, p. 118.

59. *The Problem of Asia and Its Effect Upon International Policies* (1900), p. 40.

Speed

"The true speed of war is not headlong precipitancy, but the unremitting energy that wastes no time."[60]

Strategy

"In a sea war, as in all others, two things are from the first essential—a suitable base upon the frontier, in this case the seaboard, from which the operations start, and an organized military force, in this case a fleet, of size and quality adequate to the proposed operations. If the war, as in the present instance [War of the American Revolution], extends to distance parts of the globe, there will be needed in each of those distant regions secure ports for the shipping to serve as secondary, or contingent, bases of the local war. Between those secondary and the principle, or home, bases there must be reasonably secure communication, which will depend upon military control of the intervening sea. This control must be exercised by the navy, which will enforce it by either clearing the sea in all directions of hostile cruisers, thus allowing the ships of its own nation to pass with reasonable security, or by accompanying in force [convoying] each train of supply-ships necessary for the support of the distant operation."[61]

Study

"Men who deliberately postpone the formation of opinion until the day of action, who expect from a moment of inspiration the results commonly obtained only from study and reflection, who hope for victory in ignorance of the rules that have generally given victory, are guilty of a yet greater folly, for they disregard all the past experience of our race . . .

60. *Lessons of the War With Spain,* p. 83.

61. *The Influence of Sea Power Upon History, 1660–1783* (1957 ed.), p. 460.

'Upon the field of battle,' says the great Napoleon, 'the happiest inspiration is most often only a recollection.'"[62]

Submarines

"The submarine, as so far developed, possesses particular value only in the cases where the fleet to which it belongs is not exposed; for when this comes out into the open it meets the enemy's submarines. In itself a new invention, it is but a step—though a more important one—in the progression of torpedo warfare."[63]

Trade

"The English and Dutch were no less desirous of gain than the southern nations. Each in turn has been called 'a nation of shopkeepers;' but the jeer, in so far as it is just, is to the credit of their wisdom and uprightness . . . they were more patient, in that they sought riches not by the sword but by labor, which is the reproach meant to be implied by the epithet; for thus they took the longest, instead of what seemed the shortest, road to wealth."[64]

Uncertainty and Chance

"Napoleon once said that the art of war consists in getting the most chances in your own favor."[65]

War

"War is violence, wounds, and death. Needless bloodshed is to be avoided; but even more, at the present day, is to be deprecated the view that the objects of a war are to be sacrificed to the preservation of life."[66]

62. *Naval Strategy*, p. 300.
63. "The Submarine and Its Enemies," *Collier's Weekly*, 6 April 1907.
64. *The Influence of Sea Power Upon History, 1660–1783* p. 52.
65. *Naval Strategy*, p. 177.
66. *Sea Power in Relation to the War of 1812*, vol. 2, p. 403.

"From Jomini also I imbibed a fixed disbelief in the thoughtlessly accepted maxim that the statesman and general occupy unrelated fields. For this misconception I substituted a tenet of my own, that war is simply a violent political movement; and from an expression of his, 'The sterile glory of fighting battles merely to win them,' I deduced, what military men are prone to overlook, that 'War is not fighting, but business.' "[67]

Warships

"You cannot have everything. If you attempt it, you will lose everything. On a given tonnage . . . there cannot be the highest speed *and* the thickest armor, *and* the heaviest battery, *and* the longest coal endurance."[68]

SUMMARY

Mahan would not have liked the thought that some historian of naval thought might one day tear apart his books and reassemble whole chapters together in another form, accompanied by a selection of quotations illustrating his ideas. Such an assemblage might strike him as an attempt to create aphorisms with his words or provide rules to be blindly followed by naval officers. He would have much preferred the reprinting of all his works, replete with numerous examples from history, and showing the variety of nuance involved in each application of a strategic principle. Yet, for our purposes through this series we are trying to show the nature and character of Mahan's ideas within the context of the general development of naval thought. Such an anthology, however repugnant to the author's own approach, is a practical necessity.

67. *From Sail to Steam: Recollections of a Naval Life* (1901), p. 283.
68. *Naval Strategy*, p. 44.

Here one can find, in a single volume suitable for use in the classroom or in a scholar's study, the essential points of Mahan's philosophy. Through this collection, one can ascertain clearly the fundamental focus of his work and to see its place as the initial stimulus to the development of modern naval thought. At the same time, one can see both the strengths and weaknesses of his contribution, and in comparison with other volumes in this series, define more carefully his place in the evolution of naval thought. Here one can see the initial attempt to form naval theory by linking it with military theory. Most clearly, it is an opening effort: unrefined, tied too closely to the particular outlook of Mahan's own time, and limited in its validity as theory. Nevertheless, even with those serious faults, Mahan's contribution was brilliantly insightful and a critical part of the navy's intellectual heritage.

A century after the first publication of *The Influence of Sea Power Upon History,* one can say of Mahan's work that it formed the most powerful single influence on the formulation of naval thought. Sir Julian Corbett extracted from it the basic ideas and went on to refine them into the most complete statement of classical naval theory, through further historical research and consideration of additional aspects of military theory. While Corbett's *Some Principles of Maritime Strategy* superseded Mahan's statement of strategy, the continued development of naval strategic theory has continued to highlight Mahan's contribution as the most powerful, initial impulse in its development. Mahan's writings remain an important part of the intellectual heritage of the navy and must be a necessary part of any historical examination of the development of naval thought.

John B. Hattendorf

Mahan on Naval Strategy

CHAPTER I

INTRODUCTORY

THE HISTORY of Sea Power is largely, though by no means solely, a narrative of contests between nations, of mutual rivalries, of violence frequently culminating in war. The profound influence of sea commerce upon the wealth and strength of countries was clearly seen long before the true principles which governed its growth and prosperity were detected. To secure to one's own people a disproportionate share of such benefits, every effort was made to exclude others, either by the peaceful legislative methods of monopoly or prohibitory regulations, or, when these failed, by direct violence. The clash of interests, the angry feelings roused by conflicting attempts thus to appropriate the larger share, if not the whole, of the advantages of commerce, and of distant unsettled commercial regions, led to wars. On the other hand, wars arising from other causes have been greatly modified in their conduct and issue by the control of the sea. Therefore the history of sea power, while embracing in its broad sweep all that tends to make people great upon the sea or by the sea, is largely a military history; and it is in this aspect that it will be mainly, though not exclusively, regarded in the following pages.

A study of the military history of the past, such as this, is

The Influence of Sea Power Upon History 1660–1783 (Boston, 1890), 1–24.

enjoined by great military leaders as essential to correct ideas and to the skilful conduct of war in the future. Napoleon names among the campaigns to be studied by the aspiring soldier, those of Alexander, Hannibal, and Cæsar, to whom gunpowder was unknown; and there is a substantial agreement among professional writers that, while many of the conditions of war vary from age to age with the progress of weapons, there are certain teachings in the school of history which remain constant, and being, therefore, of universal application, can be elevated to the rank of general principles. For the same reason the study of the sea history of the past will be found instructive, by its illustration of the general principles of maritime war, notwithstanding the great changes that have been brought about in naval weapons by the scientific advances of the past half century, and by the introduction of steam as the motive power.

It is doubly necessary thus to study critically the history and experience of naval warfare in the days of sailing-ships, because while these will be found to afford lessons of present application and value, steam navies have as yet made no history which can be quoted as decisive in its teaching. Of the one we have much experimental knowledge; of the other, practically none. Hence theories about the naval warfare of the future are almost wholly presumptive; and although the attempt has been made to give them a more solid basis by dwelling upon the resemblance between fleets of steamships and fleets of galleys moved by oars, which have a long and well-known history, it will be well not to be carried away by this analogy until it has been thoroughly tested. The resemblance is indeed far from superficial. The feature which the steamer and the galley have in common is the ability to move in any direction independent of the wind. Such a power makes a radical distinction between those classes of vessels and the sailing-ship; for the latter can follow only a limited number of courses when the wind blows, and must remain motionless when it fails. But while it is wise to observe things that are alike, it is also wise to look for things that differ; for when the imagination is carried away by the detection of points of resemblance,—one of the most pleasing of mental pursuits,—it is apt to be impatient of any divergence in its new-found parallels, and so may overlook or refuse to recognize such. Thus the galley and the steamship have in common, though unequally developed, the important characteristic mentioned, but in at least two points they differ; and in an appeal to the history of the galley for lessons as to fighting steamships, the differences as well as the likeness must be kept steadily in view, or false deductions may be made. The motive power of the galley when in use necessarily and rapidly declined, because human strength could not long maintain such exhausting efforts, and consequently tactical movements could continue but for a limited time;[1] and again, during the galley period offensive weapons were not only of short range, but were almost wholly confined to hand-to-hand encounter. These two conditions led almost necessarily to a rush upon each other, not, however, without some dexterous attempts to turn or double on the enemy, followed by a hand-to-hand *mêlée*. In such a rush and such a *mêlée* a great consensus of respectable, even eminent, naval opinion of the present day finds the necessary outcome of modern naval weapons,—a kind of Donnybrook Fair, in which, as the history of *mêlées* shows, it will be hard to know friend from foe. Whatever may prove to be the worth of this opinion, it cannot claim an historical basis in the sole fact that galley and steamship can

1. Thus Hermocrates of Syracuse, advocating the policy of thwarting the Athenian expedition against his city (B.C. 413) by going boldly to meet it, and keeping on the flank of its line of advance, said: "As their advance must be slow, we shall have a thousand opportunities to attack them; but if they clear their ships for action and in a body bear down expeditiously upon us, they must ply hard at their oars, and *when spent with toil* we can fall upon them."

move at any moment directly upon the enemy, and carry a beak upon their prow, regardless of the points in which galley and steamship differ. As yet this opinion is only a presumption, upon which final judgment may well be deferred until the trial of battle has given further light. Until that time there is room for the opposite view,—that a *mêlée* between numerically equal fleets, in which skill is reduced to a minimum, is not the best that can be done with the elaborate and mighty weapons of this age. The surer of himself an admiral is, the finer the tactical development of his fleet, the better his captains, the more reluctant must he necessarily be to enter into a *mêlée* with equal forces, in which all these advantages will be thrown away, chance reign supreme, and his fleet be placed on terms of equality with an assemblage of ships which have never before acted together.[2] History has lessons as to when *mêlées* are, or are not, in order.

The galley, then, has one striking resemblance to the steamer, but differs in other important features which are not so immediately apparent and are therefore less accounted of. In the sailing-ship, on the contrary, the striking feature is the difference between it and the more modern vessel; the points of resemblance, though existing and easy to find, are not so obvious, and therefore are less heeded. This impression is enhanced by the sense of utter weakness in the sailing-ship as compared with the steamer, owing to its dependence upon the wind; forgetting that, as the former fought with its equals, the tactical lessons are valid. The galley was never reduced to impotence by a calm, and hence receives more respect in our day

2. The writer must guard himself from appearing to advocate elaborate tactical movements issuing in barren demonstrations. He believes that a fleet seeking a decisive result must close with its enemy, but not until some advantage has been obtained for the collision, which will usually be gained by manœuvring, and will fall to the best drilled and managed fleet. In truth, barren results have as often followed upon headlong, close encounters as upon the most timid tactical trifling.

than the sailing-ship; yet the latter displaced it and remained supreme until the utilization of steam. The powers to injure an enemy from a great distance, to manœuvre for an unlimited length of time without wearing out the men, to devote the greater part of the crew to the offensive weapons instead of to the oar, are common to the sailing vessel and the steamer, and are at least as important, tactically considered, as the power of the galley to move in a calm or against the wind.

In tracing resemblances there is a tendency not only to overlook points of difference, but to exaggerate points of likeness,—to be fanciful. It may be so considered to point out that as the sailing-ship had guns of long range, with comparatively great penetrative power, and carronades, which were of shorter range but great smashing effect, so the modern steamer has its batteries of long-range guns and of torpedoes, the latter being effective only within a limited distance and then injuring by smashing, while the gun, as of old, aims at penetration. Yet these are distinctly tactical considerations, which must affect the plans of admirals and captains; and the analogy is real, not forced. So also both the sailing-ship and the steamer contemplate direct contact with an enemy's vessel,—the former to carry her by boarding, the latter to sink her by ramming; and to both this is the most difficult of their tasks, for to effect it the ship must be carried to a single point of the field of action, whereas projectile weapons may be used from many points of a wide area.

The relative positions of two sailing-ships, or fleets, with reference to the direction of the wind involved most important tactical questions, and were perhaps the chief care of the seamen of that age. To a superficial glance it may appear that since this has become a matter of such indifference to the steamer, no analogies to it are to be found in present conditions, and the lessons of history in this respect are valueless. A more careful consideration of the distinguishing characteristics of the lee and the weather "gage,"[3] directed to their essential features and disregarding secondary details, will show that this is a mistake. The distinguishing feature of the weather-gage was that it conferred the power of giving or refusing battle at will, which in turn carries the usual advantage of an offensive attitude in the choice of the method of attack. This advantage was accompanied by certain drawbacks, such as irregularity introduced into the order, exposure to raking or enfilading cannonade, and the sacrifice of part or all of the artillery-fire of the assailant,—all which were incurred in approaching the enemy. The ship, or fleet, with the lee-gage could not attack; if it did not wish to retreat, its action was confined to the defensive, and to receiving battle on the enemy's terms. This disadvantage was compensated by the comparative ease of maintaining the order of battle undisturbed, and by a sustained artillery-fire to which the enemy for a time was unable to reply. Historically, these favorable and unfavorable characteristics have their counterpart and analogy in the offensive and defensive operations of all ages. The offence undertakes certain risks and disadvantages in order to reach and destroy the enemy; the defence, so long as it remains such, refuses the risks of advance, holds on to a careful, well-ordered position, and avails itself of the exposure to which the assailant submits himself. These radical differences between the weather and the lee gage were so clearly recognized, through the cloud of lesser details accompanying them, that the former was ordinarily chosen by the English, because

3. A ship was said to have the weather-gage, or "the advantage of the wind," or "to be to windward," when the wind allowed her to steer for her opponent, and did not let the latter head straight for her. The extreme case was when the wind blew direct from one to the other; but there was a large space on either side of this line to which the term "weather-gage" applied. If the lee ship be taken as the centre of a circle, there were nearly three eighths of its area in which the other might be and still keep the advantage of the wind to a greater or less degree. Lee is the opposite of weather.

their steady policy was to assail and destroy their enemy; whereas the French sought the lee-gage, because by so doing they were usually able to cripple the enemy as he approached, and thus evade decisive encounters and preserve their ships. The French, with rare exceptions, subordinated the action of the navy to other military considerations, grudged the money spent upon it, and therefore sought to economize their fleet by assuming a defensive position and limiting its efforts to the repelling of assaults. For this course the lee-gage, skilfully used, was admirably adapted so long as an enemy displayed more courage than conduct; but when Rodney showed an intention to use the advantage of the wind, not merely to attack, but to make a formidable concentration on a part of the enemy's line, his wary opponent, De Guichen, changed his tactics. In the first of their three actions the Frenchman took the lee-gage; but after recognizing Rodney's purpose he manœuvred for the advantage of the wind, not to attack, but to refuse action except on his own terms. The power to assume the offensive, or to refuse battle, rests no longer with the wind, but with the party which has the greater speed; which in a fleet will depend not only upon the speed of the individual ships, but also upon their tactical uniformity of action. Henceforth the ships which have the greatest speed will have the weather-gage.

It is not therefore a vain expectation, as many think, to look for useful lessons in the history of sailing-ships as well as in that of galleys. Both have their points of resemblance to the modern ship; both have also points of essential difference, which make it impossible to cite their experiences or modes of action as tactical *precedents* to be followed. But a precedent is different from and less valuable than a principle. The former may be originally faulty, or may cease to apply through change of circumstances; the latter has its root in the essential nature of things, and, however various its application as conditions change, remains a standard to which action must conform to attain success. War has such principles; their existence is detected by the study of the past, which reveals them in successes and in failures, the same from age to age. Conditions and weapons change; but to cope with the one or successfully wield the others, respect must be had to these constant teachings of history in the tactics of the battlefield, or in those wider operations of war which are comprised under the name of strategy.

It is however in these wider operations, which embrace a whole theatre of war, and in a maritime contest may cover a large portion of the globe, that the teachings of history have a more evident and permanent value, because the conditions remain more permanent. The theatre of war may be larger or smaller, its difficulties more or less pronounced, the contending armies more or less great, the necessary movements more or less easy, but these are simply differences of scale, of degree, not of kind. As a wilderness gives place to civilization, as means of communication multiply, as roads are opened, rivers bridged, food-resources increased, the operations of war become easier, more rapid, more extensive; but the principles to which they must be conformed remain the same. When the march on foot was replaced by carrying troops in coaches, when the latter in turn gave place to railroads, the scale of distances was increased, or, if you will, the scale of time diminished; but the principles which dictated the point at which the army should be concentrated, the direction in which it should move, the part of the enemy's position which it should assail, the protection of communications, were not altered. So, on the sea, the advance from the galley timidly creeping from port to port to the sailing-ship launching out boldly to the ends of the earth, and from the latter to the steamship of our own time, has increased the scope and the rapidity of naval operations without necessarily changing the principles which should direct them; and the speech of Hermocrates twenty-three hundred years ago, before quoted, contained a correct strategic plan, which is as applicable in its principles now as it was then. Before hostile armies or fleets are brought into con*tact* (a word which perhaps better than any other indicates the dividing line between tactics and strategy), there are a number of questions to be decided, covering the whole plan of operations throughout the theatre of war. Among these are the proper function of the navy in the war; its true objective; the point or points upon which it should be concentrated; the establishment of depots of coal and supplies; the maintenance of communications between these depots and the home base; the military value of commerce-destroying as a decisive or a secondary operation of war; the system upon which commerce-destroying can be most efficiently conducted, whether by scattered cruisers or by holding in force some vital centre through which commercial shipping must pass. All these are strategic questions, and upon all these history has a great deal to say. There has been of late a valuable discussion in English naval circles as to the comparative merits of the policies of two great English admirals, Lord Howe and Lord St. Vincent, in the disposition of the English navy when at war with France. The question is purely strategic, and is not of mere historical interest; it is of vital importance now, and the principles upon which its decision rests are the same now as then. St. Vincent's policy saved England from invasion, and in the hands of Nelson and his brother admirals led straight up to Trafalgar.

It is then particularly in the field of naval strategy that the teachings of the past have a value which is in no degree lessened. They are there useful not only as illustrative of principles, but also as precedents, owing to the comparative permanence of the conditions. This is less obviously true as to tactics, when the fleets come into collision at the point to which strategic considerations have brought them. The unresting progress of mankind causes continual change in the weapons; and with that must come a continual change in the manner of fighting,—in the handling and disposition of troops or ships on the battlefield. Hence arises a tendency on the part of many connected with maritime matters to think that no advantage is to be gained from the study of former experiences; that time so used is wasted. This view, though natural, not only leaves wholly out of sight those broad strategic considerations which lead nations to put fleets afloat, which direct the sphere of their action, and so have modified and will continue to modify the history of the world, but is one-sided and narrow even as to tactics. The battles of the past succeeded or failed according as they were fought in conformity with the principles of war; and the seaman who carefully studies the causes of success or failure will not only detect and gradually assimilate these principles, but will also acquire increased aptitude in applying them to the tactical use of the ships and weapons of his own day. He will observe also that changes of tactics have not only taken place *after* changes in weapons, which necessarily is the case, but that the interval between such changes has been unduly long. This doubtless arises from the fact that an improvement of weapons is due to the energy of one or two men, while changes in tactics have to overcome the inertia of a conservative class; but it is a great evil. It can be remedied only by a candid recognition of each change, by careful study of the powers and limitations of the new ship or weapon, and by a consequent adaptation of the method of using it to the qualities it possesses, which will constitute its tactics. History shows that it is vain to hope that military men generally will be at the pains to do this, but that the one who does will go into battle with a great advantage,—a lesson in itself of no mean value.

We may therefore accept now the words of a French tactician, Morogues, who wrote a century and a quarter ago: "Naval tactics are based upon conditions the chief causes of which, namely the arms, may change; which in turn causes necessarily a change in the construction of ships, in the manner of handling them, and so finally in the disposition and handling of fleets." His further statement, that "it is not a science founded upon principles absolutely invariable," is more open to criticism. It would be more correct to say that the application of its principles varies as the weapons change. The application of the principles doubtless varies also in strategy from time to time, but the variation is far less; and hence the recognition of the underlying principle is easier. This statement is of sufficient importance to our subject to receive some illustrations from historical events.

The battle of the Nile, in 1798, was not only an overwhelming victory for the English over the French fleet, but had also the decisive effect of destroying the communications between France and Napoleon's army in Egypt. In the battle itself the English admiral, Nelson, gave a most brilliant example of grand tactics, if that be, as has been defined, "the art of making good combinations preliminary to battles as well as during their progress." The particular tactical combination depended upon a condition now passed away, which was the inability of the lee ships of a fleet at anchor to come to the help of the weather ones before the latter were destroyed; but the principles which underlay the combination, namely, to choose that part of the enemy's order which can least easily be helped, and to attack it with superior forces, has not passed away. The action of Admiral Jervis at Cape St. Vincent, when with fifteen ships he won a victory over twenty-seven, was dictated by the same principle, though in this case the enemy was not at anchor, but under way. Yet men's minds are so constituted that they seem more impressed by the transiency of the conditions than by the undying principle which coped with them. In the strategic effect of Nelson's victory upon the course of the war, on the contrary, the principle involved is not only more easily recognized, but it is at once seen to be applicable to our own day. The issue of the enterprise in Egypt depended upon keeping open the communications with

France. The victory of the Nile destroyed the naval force, by which alone the communications could be assured, and determined the final failure; and it is at once seen, not only that the blow was struck in accordance with the principle of striking at the enemy's line of communication, but also that the same principle is valid now, and would be equally so in the days of the galley as of the sailing-ship or steamer.

Nevertheless, a vague feeling of contempt for the past, supposed to be obsolete, combines with natural indolence to blind men even to those permanent strategic lessons which lie close to the surface of naval history. For instance, how many look upon the battle of Trafalgar, the crown of Nelson's glory and the seal of his genius, as other than an isolated event of exceptional grandeur? How many ask themselves the strategic question, "How did the ships come to be just there?" How many realize it to be the final act in a great strategic drama, extending over a year or more, in which two of the greatest leaders that ever lived, Napoleon and Nelson, were pitted against each other? At Trafalgar it was not Villeneuve that failed, but Napoleon that was vanquished; not Nelson that won, but England that was saved; and why? Because Napoleon's combinations failed, and Nelson's intuitions and activity kept the English fleet ever on the track of the enemy, and brought it up in time at the decisive moment.[4] The tactics at Trafalgar, while open to criticism in detail, were in their main features conformable to the principles of war, and their audacity was justified as well by the urgency of the case as by the results; but the great lessons of efficiency in preparation, of activity and energy in execution, and of thought and insight on the part of the English leader during the previous months, are strategic lessons, and as such they still remain good.

4. See note at end of this chapter, pp. 24–26.

In these two cases events were worked out to their natural and decisive end. A third may be cited, in which, as no such definite end was reached, an opinion as to what should have been done may be open to dispute. In the war of the American Revolution, France and Spain became allies against England in 1779. The united fleets thrice appeared in the English Channel, once to the number of sixty-six sail of the line, driving the English fleet to seek refuge in its ports because far inferior in numbers. Now, the great aim of Spain was to recover Gibraltar and Jamaica; and to the former end immense efforts both by land and sea were put forth by the allies against that nearly impregnable fortress. They were fruitless. The question suggested—and it is purely one of naval strategy—is this: "Would not Gibraltar have been more surely recovered by controlling the English Channel, attacking the British fleet even in its harbors, and threatening England with annihilation of commerce and invasion at home, than by far greater efforts directed against a distant and very strong outpost of her empire? The English people, from long immunity, were particularly sensitive to fears of invasion, and their great confidence in their fleets, if rudely shaken, would have left them proportionately disheartened. However decided, the question as a point of strategy is fair; and it is proposed in another form by a French officer of the period, who favored directing the great effort on a West India island which might be exchanged against Gibraltar. It is not, however, likely that England would have given up the key of the Mediterranean for any other foreign possession, though she might have yielded to save her firesides and her capital. Napoleon once said that he would reconquer Pondicherry on the banks of the Vistula. Could he have controlled the English Channel, as the allied fleet did for a moment in 1779, can it be doubted that he would have conquered Gibraltar on the shores of England?

To impress more strongly the truth that history both suggests strategic study and illustrates the principles of war by the facts which it transmits, two more instances will be taken, which are more remote in time than the period specially considered in this work. How did it happen that, in two great contests between the powers of the East and of the West in the Mediterranean, in one of which the empire of the known world was at stake, the opposing fleets met on spots so near each other as Actium and Lepanto? Was this a mere coincidence, or was it due to conditions that recurred, and may recur again?[5] If the latter, it is worth while to study out the reason; for if there should again arise a great eastern power of the sea like that of Antony or of Turkey, the strategic questions would be similar. At present, indeed, it seems that the centre of sea power, resting mainly with England and France, is overwhelmingly in the West; but should any chance add to the control of the Black Sea basin, which Russia now has, the possession of the entrance to the Mediterranean, the existing strategic conditions affecting sea power would all be modified. Now, were the West arrayed against the East, England and France would go at once unopposed to the Levant, as they did in 1854, and as England alone went in 1878; in case of the change suggested, the East, as twice before, would meet the West half-way.

At a very conspicuous and momentous period of the world's history, Sea Power had a strategic bearing and weight which has received scant recognition. There cannot now be had the full knowledge necessary for tracing in detail its influence upon the issue of the second Punic War; but the indications which remain are sufficient to warrant the assertion that it was a determining factor. An accurate judgment upon this point cannot be formed by mastering only such facts of the particular contest as have been clearly transmitted, for as

5. The battle of Navarino (1827) between Turkey and the Western Powers was fought in this neighborhood.

usual the naval transactions have been slightingly passed over; there is needed also familiarity with the details of general naval history in order to draw, from slight indications, correct inferences based upon a knowledge of what has been possible at periods whose history is well known. The control of the sea, however real, does not imply that an enemy's single ships or small squadrons cannot steal out of port, cannot cross more or less frequented tracts of ocean, make harassing descents upon unprotected points of a long coastline, enter blockaded harbors. On the contrary, history has shown that such evasions are always possible, to some extent, to the weaker party, however great the inequality of naval strength. It is not therefore inconsistent with the general control of the sea, or of a decisive part of it, by the Roman fleets, that the Carthaginian admiral Bomilcar in the fourth year of the war, after the stunning defeat of Cannæ, landed four thousand men and a body of elephants in south Italy; nor that in the seventh year, flying from the Roman fleet off Syracuse, he again appeared at Tarentum, then in Hannibal's hands; nor that Hannibal sent despatch vessels to Carthage; nor even that, at last, he withdrew in safety to Africa with his wasted army. None of these things prove that the government in Carthage could, if it wished, have sent Hannibal the constant support which, as a matter of fact, he did not receive; but they do tend to create a natural impression that such help could have been given. Therefore the statement, that the Roman preponderance at sea had a decisive effect upon the course of the war, needs to be made good by an examination of ascertained facts. Thus the kind and degree of its influence may be fairly estimated.

At the beginning of the war, Mommsen says, Rome controlled the seas. To whatever cause, or combination of causes, it be attributed, this essentially non-maritime state had in the first Punic War established over its sea-faring rival a naval supremacy, which still lasted. In the second war there was no naval battle of importance,—a circumstance which in itself, and still more in connection with other well-ascertained facts, indicates a superiority analogous to that which at other epochs has been marked by the same feature.

As Hannibal left no memoirs, the motives are unknown which determined him to the perilous and almost ruinous march through Gaul and across the Alps. It is certain, however, that his fleet on the coast of Spain was not strong enough to contend with that of Rome. Had it been, he might still have followed the road he actually did, for reasons that weighed with him; but had he gone by the sea, he would not have lost thirty-three thousand out of sixty thousand veteran soldiers with whom he started.

While Hannibal was making this dangerous march, the Romans were sending to Spain, under the two elder Scipios, one part of their fleet, carrying a consular army. This made the voyage without serious loss, and the army established itself successfully north of the Ebro, on Hannibal's line of communications. At the same time another squadron, with an army commanded by the other consul, was sent to Sicily. The two together numbered two hundred and twenty ships. On its station each met and defeated a Carthaginian squadron with an ease which may be inferred from the slight mention made of the actions, and which indicates the actual superiority of the Roman fleet.

After the second year the war assumed the following shape: Hannibal, having entered Italy by the north, after a series of successes had passed southward around Rome and fixed himself in southern Italy, living off the country,—a condition which tended to alienate the people, and was especially precarious when in contact with the mighty political and military system of control which Rome had there established. It was therefore from the first urgently necessary that he should establish, between himself and some reliable base, that stream of supplies and reinforcements which in terms of modern war is called "communications." There were three friendly regions which might, each or all, serve as such a base,—Carthage itself, Macedonia, and Spain. With the first two, communication could be had only by sea. From Spain, where his firmest support was found, he could be reached by both land and sea, unless an enemy barred the passage; but the sea route was the shorter and easier.

In the first years of the war, Rome, by her sea power, controlled absolutely the basin between Italy, Sicily, and Spain, known as the Tyrrhenian and Sardinian Seas. The seacoast from the Ebro to the Tiber was mostly friendly to her. In the fourth year, after the battle of Cannæ, Syracuse forsook the Roman alliance, the revolt spread through Sicily, and Macedonia also entered into an offensive league with Hannibal. These changes extended the necessary operations of the Roman fleet, and taxed its strength. What disposition was made of it, and how did it thereafter influence the struggle?

The indications are clear that Rome at no time ceased to control the Tyrrhenian Sea, for her squadrons passed unmolested from Italy to Spain. On the Spanish coast also she had full sway till the younger Scipio saw fit to lay up the fleet. In the Adriatic, a squadron and naval station were established at Brindisi to check Macedonia, which performed their task so well that not a soldier of the phalanxes ever set foot in Italy. "The want of a war fleet," says Mommsen, "paralyzed Philip in all his movements." Here the effect of Sea Power is not even a matter of inference.

In Sicily, the struggle centred about Syracuse. The fleets of Carthage and Rome met there, but the superiority evidently lay with the latter; for though the Carthaginians at times succeeded in throwing supplies into the city, they avoided meeting the Roman fleet in battle. With Lilybæum, Palermo, and Messina in its hands, the latter was well based in the north coast of the island. Access by the south was left open to the Carthaginians, and they were thus able to maintain the insurrection.

Putting these facts together, it is a reasonable inference, and supported by the whole tenor of the history, that the Roman sea power controlled the sea north of a line drawn from Tarragona in Spain to Lilybæum (the modern Marsala), at the west end of Sicily, thence round by the north side of the island through the straits of Messina down to Syracuse, and from there to Brindisi in the Adriatic. This control lasted, unshaken, throughout the war. It did not exclude maritime raids, large or small, such as have been spoken of; but it did forbid the sustained and secure communications of which Hannibal was in deadly need.

On the other hand, it seems equally plain that for the first ten years of the war the Roman fleet was not strong enough for sustained operations in the sea between Sicily and Carthage, nor indeed much to the south of the line indicated. When Hannibal started, he assigned such ships as he had to maintaining the communications between Spain and Africa, which the Romans did not then attempt to disturb.

The Roman sea power, therefore, threw Macedonia wholly out of the war. It did not keep Carthage from maintaining a useful and most harassing diversion in Sicily; but it did prevent her sending troops, when they would have been most useful, to her great general in Italy. How was it as to Spain?

Spain was the region upon which the father of Hannibal and Hannibal himself had based their intended invasion of Italy. For eighteen years before this began they had occupied the country, extending and consolidating their power, both political and military, with rare sagacity. They had raised, and trained in local wars, a large and now veteran army. Upon his own departure, Hannibal intrusted the government to his younger brother, Hasdrubal, who preserved toward him to the end a loyalty and devotion which he had no reason to hope from the faction-cursed mother-city in Africa.

At the time of his starting, the Carthaginian power in Spain was secured from Cadiz to the river Ebro. The region between this river and the Pyrenees was inhabited by tribes friendly to the Romans, but unable, in the absence of the latter, to oppose a successful resistance to Hannibal. He put them down, leaving eleven thousand soldiers under Hanno to keep military possession of the country, lest the Romans should establish themselves there, and thus disturb his communications with his base.

Cnæus Scipio, however, arrived on the spot by sea the same year with twenty thousand men, defeated Hanno, and occupied both the coast and interior north of the Ebro. The Romans thus held ground by which they entirely closed the road between Hannibal and reinforcements from Hasdrubal, and whence they could attack the Carthaginian power in Spain; while their own communications with Italy, being by water, were secured by their naval supremacy. They made a naval base at Tarragona, confronting that of Hasdrubal at Cartagena, and then invaded the Carthaginian dominions. The war in Spain went on under the elder Scipios, seemingly a side issue, with varying fortune for seven years; at the end of which time Hasdrubal inflicted upon them a crushing defeat, the two brothers were killed, and the Carthaginians nearly succeeded in breaking through to the Pyrenees with reinforcements for Hannibal. The attempt, however, was checked for the moment; and before it could be renewed, the fall of Capua released twelve thousand veteran Romans, who were sent to Spain under Claudius Nero, a man of exceptional ability, to whom was due later the most decisive military movement made by any Roman general during the Second Punic War. This seasonable reinforcement, which again assured the shaken grip on Hasdrubal's line of march, came by sea,—a way which, though most rapid and easy, was closed to the Carthaginians by the Roman navy.

Two years later the younger Publius Scipio, celebrated afterward as Africanus, received the command in Spain, and captured Cartagena by a combined military and naval attack; after which he took the most extraordinary step of breaking up his fleet and transferring the seamen to the army. Not contented to act merely as the "containing"[6] force against Hasdrubal by closing the passes of the Pyrenees, Scipio pushed forward into southern Spain, and fought a severe but indecisive battle on the Guadalquivir; after which Hasdrubal slipped away from him, hurried north, crossed the Pyrenees at their extreme west, and pressed on to Italy, where Hannibal's position was daily growing weaker, the natural waste of his army not being replaced.

The war lasted ten years, when Hasdrubal, having met little loss on the way, entered Italy at the north. The troops he brought, could they be safely united with those under the command of the unrivalled Hannibal, might give a decisive turn to the war, for Rome herself was nearly exhausted; the iron links which bound her own colonies and the allied States to her were strained to the utmost, and some had already snapped. But the military position of the two brothers was also perilous in the extreme. One being at the river Metaurus, the other in Apulia, two hundred miles apart, each was confronted by a superior enemy, and both these Roman armies were between their separated opponents. This false situation, as well as the long delay of Hasdrubal's coming, was due to the Roman control of the sea, which throughout the war limited the mutual support of the Carthaginian brothers to the route through Gaul. At the very time that Hasdrubal was

6. A "containing" force is one to which, in a military combination, is assigned the duty of stopping, or delaying the advance of a portion of the enemy, while the main effort of the army or armies is being exerted in a different quarter.

making his long and dangerous circuit by land, Scipio had sent eleven thousand men from Spain by sea to reinforce the army opposed to him. The upshot was that messengers from Hasdrubal to Hannibal, having to pass over so wide a belt of hostile country, fell into the hands of Claudius Nero, commanding the southern Roman army, who thus learned the route which Hasdrubal intended to take. Nero correctly appreciated the situation, and, escaping the vigilance of Hannibal, made a rapid march with eight thousand of his best troops to join the forces in the north. The junction being effected, the two consuls fell upon Hasdrubal in overwhelming numbers and destroyed his army; the Carthaginian leader himself falling in the battle. Hannibal's first news of the disaster was by the head of his brother being thrown into his camp. He is said to have exclaimed that Rome would now be mistress of the world; and the battle of Metaurus is generally accepted as decisive of the struggle between the two States.

The military situation which finally resulted in the battle of the Metaurus and the triumph of Rome may be summed up as follows: To overthrow Rome it was necessary to attack her in Italy at the heart of her power, and shatter the strongly linked confederacy of which she was the head. This was the objective. To reach it, the Carthaginians needed a solid base of operations and a secure line of communications. The former was established in Spain by the genius of the great Barca family; the latter was never achieved. There were two lines possible,—the one direct by sea, the other circuitous through Gaul. The first was blocked by the Roman sea power, the second imperilled and finally intercepted through the occupation of northern Spain by the Roman army. This occupation was made possible through the control of the sea, which the Carthaginians never endangered. With respect to Hannibal and his base, therefore, Rome occupied two central positions, Rome itself and northern Spain, joined by an easy interior line of communications, the sea; by which mutual support was continually given.

Had the Mediterranean been a level desert of land, in which the Romans held strong mountain ranges in Corsica and Sardinia, fortified posts at Tarragona, Lilybæum, and Messina, the Italian coast-line nearly to Genoa, and allied fortresses in Marseilles and other points; had they also possessed an armed force capable by its character of traversing that desert at will, but in which their opponents were very inferior and therefore compelled to a great circuit in order to concentrate their troops, the military situation would have been at once recognized, and no words would have been too strong to express the value and effect of that peculiar force. It would have been perceived, also, that the enemy's force of the same kind might, however inferior in strength, make an inroad, or raid, upon the territory thus held, might burn a village or waste a few miles of borderland, might even cut off a convoy at times, without, in a military sense, endangering the communications. Such predatory operations have been carried on in all ages by the weaker maritime belligerent, but they by no means warrant the inference, irreconcilable with the known facts, "that neither Rome nor Carthage could be said to have undisputed mastery of the sea," because "Roman fleets sometimes visited the coasts of Africa, and Carthaginian fleets in the same way appeared off the coast of Italy." In the case under consideration, the navy played the part of such a force upon the supposed desert; but as it acts on an element strange to most writers, as its members have been from time immemorial a strange race apart, without prophets of their own, neither themselves nor their calling understood, its immense determining influence upon the history of that era, and consequently upon the history of the world, has been overlooked. If the preceding argument is sound, it is as defective to omit sea power from the list of principal factors in the result, as it would be absurd to claim for it an exclusive influence.

Instances such as have been cited, drawn from widely separated periods of time, both before and after that specially treated in this work, serve to illustrate the intrinsic interest of the subject, and the character of the lessons which history has to teach. As before observed, these come more often under the head of strategy than of tactics; they bear rather upon the conduct of campaigns than of battles, and hence are fraught with more lasting value. To quote a great authority in this connection, Jomini says: "Happening to be in Paris near the end of 1851, a distinguished person did me the honor to ask my opinion as to whether recent improvements in fire arms would cause any great modifications in the way of making war. I replied that they would probably have an influence upon the details of tactics, but that in great strategic operations and the grand combinations of battles, victory would, now as ever, result from the application of the principles which had led to the success of great generals in all ages; of Alexander and Cæsar, as well as of Frederick and Napoleon." This study has become more than ever important now to navies, because of the great and steady power of movement possessed by the modern steamer. The best-planned schemes might fail through stress of weather in the days of the galley and the sailing-ship; but this difficulty has almost disappeared. The principles which should direct great naval combinations have been applicable to all ages, and are deducible from history; but the power to carry them out with little regard to the weather is a recent gain.

The definitions usually given of the word "strategy" confine it to military combinations embracing one or more fields of operations, either wholly distinct or mutually dependent, but always regarded as actual or immediate scenes of war. However this may be on shore, a recent French author is quite right in pointing out that such a definition is too narrow for naval strategy. "This," he says, "differs from military strategy in that it is as necessary in peace as in war. Indeed, in peace it may gain its most decisive victories by occupying in a country, either by purchase or treaty, excellent positions which would perhaps hardly be got by war. It learns to profit by all opportunities of settling on some chosen point of a coast, and to render definitive an occupation which at first was only transient." A generation that has seen England within ten years occupy successively Cyprus and Egypt, under terms and conditions on their face transient, but which have not yet led to the abandonment of the positions taken, can readily agree with this remark; which indeed receives constant illustration from the quiet persistency with which all the great sea powers are seeking position after position, less noted and less noteworthy than Cyprus and Egypt, in the different seas to which their people and their ships penetrate. "Naval strategy has indeed for its end to found, support, and increase, as well in peace as in war, the sea power of a country;" and therefore its study has an interest and value for all citizens of a free country, but especially for those who are charged with its foreign and military relations.

The general conditions that either are essential to or powerfully affect the greatness of a nation upon the sea will now be examined; after which a more particular consideration of the various maritime nations of Europe at the middle of the seventeenth century, where the historical survey begins, will serve at once to illustrate and give precision to the conclusions upon the general subject.

Note.—The brilliancy of Nelson's fame, dimming as it does that of all his contemporaries, and the implicit trust felt by England in him as the one man able to save her from the schemes of Napoleon, should not of course obscure the fact that only one portion of the field was, or could be, occupied by him. Napoleon's aim, in the campaign which ended at Trafalgar, was to unite in the West Indies the French fleets of Brest, Toulon, and Rochefort, together with a strong body of Spanish ships, thus forming an overwhelming force which he intended should return together to the English Channel and cover the crossing of the French army. He naturally expected that, with England's interests scattered all over the world, confusion and distraction would arise from ignorance and the destination of the French squadrons, and the English navy be drawn away from his objective point. The portion of the field committed to Nelson was the Mediterranean, where he watched the great arsenal of Toulon and the highways alike to the East and to the Atlantic. This was inferior in consequence to no other, and assumed additional importance in the eyes of Nelson from his conviction that the former attempts on Egypt would be renewed. Owing to this persuasion he took at first a false step, which delayed his pursuit of the Toulon fleet when it sailed under the command of Villeneuve; and the latter was further favored by a long continuance of fair winds, while the English had head winds. But while all this is true, while the failure of Napoleon's combinations must be attributed to the tenacious grip of the English blockade off Brest, *as well as* to Nelson's energetic pursuit of the Toulon fleet when it escaped to the West Indies and again on its hasty return to Europe, the latter is fairly entitled to the eminent distinction which history has accorded it, and which is asserted in the text. Nelson did not, indeed, fathom the intentions of Napoleon. This may have been owing, as some have said, to lack of insight; but it may be more simply laid to the usual disadvantage under which the defence lies before the blow has fallen, of ignorance as to the point threatened by the offence. It is insight enough to fasten on the key of a situation; and this Nelson rightly saw was the fleet, not the station. Consequently, his action has afforded a striking instance of how tenacity of purpose and untiring energy in execution can repair a first mistake and baffle deeply laid plans. His Mediterranean command embraced many duties and cares; but amid and dominating them all, he saw clearly the Toulon fleet as the controlling factor there, and an important factor in any naval combination of the Emperor. Hence his attention was unwaveringly fixed upon it; so much so that he called it "his fleet," a phrase which has somewhat vexed the sensibilities of French critics. This simple and accurate view of the military situation strengthened him in taking the fearless resolution and bearing the immense responsibility of abandoning his station in order to follow "his fleet." Determined thus on a pursuit the undeniable wisdom of which should not obscure the greatness of mind that undertook it, he followed so vigorously as to reach Cadiz on his return a week before Villeneuve entered Ferrol, despite unavoidable delays arising from false information and uncertainty as to the enemy's movements. The same untiring ardor enabled him to bring up his own ships from Cadiz to Brest in time to make the fleet there superior to Villeneuve's, had the latter persisted in his attempt to reach the neighborhood. The English, very inferior in aggregate number of vessels to the allied fleets, were by this seasonable reinforcement of eight veteran ships put into the best possible position strategically, as will be pointed out in dealing with similar conditions in the war of the American Revolution. Their forces were united in one great fleet in the Bay of Biscay, interposed between the two divisions of the enemy in Brest and Ferrol, superior in number to either singly, and with a strong probability of being able to deal with one before the other could come up. This was due to able action all round on the part of the English authorities; but above all other factors in the result stands Nelson's single-minded pursuit of "his fleet."

This interesting series of strategic movements ended on the 14th of August, when Villeneuve, in despair of reaching Brest, headed for Cadiz, where he anchored on the 20th. As soon as Napoleon heard of this, after an outburst of rage against the admiral, he at once dictated the series of movements which resulted in Ulm and Austerlitz, abandoning his purposes against England. The battle of Trafalgar, fought October 21, was therefore separated by a space of two months from the extensive movements of which it was nevertheless the outcome. Isolated from them in point of time, it was none the less the seal of Nelson's genius, affixed later to the record he had made in the near past. With equal truth it is said that England was saved at Trafalgar, though the Emperor had then given up his intended invasion; the destruction there emphasized and sealed the strategic triumph which had noiselessly foiled Napoleon's plans.

CHAPTER II

DISCUSSION OF THE ELEMENTS OF SEA POWER

THE first and most obvious light in which the sea presents itself from the political and social point of view is that of a great highway; or better, perhaps, of a wide common, over which men may pass in all directions, but on which some well-worn paths show that controlling reasons have led them to choose certain lines of travel rather than others. These lines of travel are called trade routes; and the reasons which have determined them are to be sought in the history of the world.

Notwithstanding all the familiar and unfamiliar dangers of the sea, both travel and traffic by water have always been easier and cheaper than by land. The commercial greatness of Holland was due not only to her shipping at sea, but also to the numerous tranquil water-ways which gave such cheap and easy access to her own interior and to that of Germany. This advantage of carriage by water over that by land was yet more marked in a period when roads were few and very bad, wars frequent and society unsettled, as was the case two hundred years ago. Sea traffic then went in peril of robbers, but was nevertheless safer and quicker than that by land. A Dutch writer of that time, estimating the chances of his country in a war with England, notices among other things that the water-ways of England failed to penetrate the country sufficiently;

Chapter I, *The Influence of Sea Power Upon History* (Boston, 1890).

therefore, the roads being bad, goods from one part of the kingdom to the other must go by sea, and be exposed to capture by the way. As regards purely internal trade, this danger has generally disappeared at the present day. In most civilized countries, now, the destruction or disappearance of the coasting trade would only be an inconvenience, although water transit is still the cheaper. Nevertheless, as late as the wars of the French Republic and the First Empire, those who are familiar with the history of the period, and the light naval literature that has grown up around it, know how constant is the mention of convoys stealing from point to point along the French coast, although the sea swarmed with English cruisers and there were good inland roads.

Under modern conditions, however, home trade is but a part of the business of a country bordering on the sea. Foreign necessaries or luxuries must be brought to its ports, either in its own or in foreign ships, which will return, bearing in exchange the products of the country, whether they be the fruits of the earth or the works of men's hands; and it is the wish of every nation that this shipping business should be done by its own vessels. The ships that thus sail to and fro must have secure ports to which to return, and must, as far as possible, be followed by the protection of their country throughout the voyage.

This protection in time of war must be extended by armed shipping. The necessity of a navy, in the restricted sense of the word, springs, therefore, from the existence of a peaceful shipping, and disappears with it, except in the case of a nation which has aggressive tendencies, and keeps up a navy merely as a branch of the military establishment. As the United States has at present no aggressive purposes, and as its merchant service has disappeared, the dwindling of the armed fleet and general lack of interest in it are strictly logical consequences. When for any reason sea trade is again found to pay, a large enough shipping interest will reappear to compel the revival of the war fleet. It is possible that when a canal route through the Central-American Isthmus is seen to be a near certainty, the aggressive impulse may be strong enough to lead to the same result. This is doubtful, however, because a peaceful, gain-loving nation is not far-sighted, and far-sightedness is needed for adequate military preparation, especially in these days.

As a nation, with its unarmed and armed shipping, launches forth from its own shores, the need is soon felt of points upon which the ships can rely for peaceful trading, for refuge and supplies. In the present day friendly, though foreign, ports are to be found all over the world; and their shelter is enough while peace prevails. It was not always so, nor does peace always endure, though the United States have been favored by so long a continuance of it. In earlier times the merchant seaman, seeking for trade in new and unexplored regions, made his gains at risk of life and liberty from suspicious or hostile nations, and was under great delays in collecting a full and profitable freight. He therefore intuitively sought at the far end of his trade route one or more stations, to be given to him by force or favor, where he could fix himself or his agents in reasonable security, where his ships could lie in safety, and where the merchantable products of the land could be continually collecting, awaiting the arrival of the home fleet, which should carry them to the mother-country. As there was immense gain, as well as much risk, in these early voyages, such establishments naturally multiplied and grew until they became colonies; whose ultimate development and success depended upon the genius and policy of the nation from which they sprang, and form a very great part of the history, and particularly of the sea history, of the world. All colonies had not the simple and natural birth and growth above described. Many were more formal, and purely political, in their conception and founding, the act of the rulers of the people rather than of private individuals; but the trading-station with its after expansion, the work simply of the adventurer seeking gain, was in its reasons and essence the same as the elaborately organized and chartered colony. In both cases the mother-country had won a foothold in a foreign land, seeking a new outlet for what it had to sell, a new sphere for its shipping, more employment for its people, more comfort and wealth for itself.

The needs of commerce, however, were not all provided for when safety had been secured at the far end of the road. The voyages were long and dangerous, the seas often beset with enemies. In the most active days of colonizing there prevailed on the sea a lawlessness the very memory of which is now almost lost, and the days of settled peace between maritime nations were few and far between. Thus arose the demand for stations along the road, like the Cape of Good Hope, St. Helena, and Mauritius, not primarily for trade, but for defence and war; the demand for the possession of posts like Gibraltar, Malta, Louisburg, at the entrance of the Gulf of St. Lawrence,—posts whose value was chiefly strategic, though not necessarily wholly so. Colonies and colonial posts were sometimes commercial, sometimes military in their character; and it was exceptional that the same position was equally important in both points of view, as New York was.

In these three things—production, with the necessity of exchanging products, shipping, whereby the exchange is carried on, and colonies, which facilitate and enlarge the operations of shipping and tend to protect it by multiplying points of safety—is to be found the key to much of the history, as well as of the policy, of nations bordering upon the sea. The policy has varied both with the spirit of the age and with the character and clear-sightedness of the rulers; but the history of the seaboard nations has been less determined by the shrewdness and foresight of governments than by conditions of position, extent, configuration, number and character of their people,—by what are called, in a word, natural conditions. It must however be admitted, and will be seen, that the wise or unwise action of individual men has at certain periods had a great modifying influence upon the growth of sea power in the broad sense, which includes not only the military strength afloat, that rules the sea or any part of it by force of arms, but also the peaceful commerce and shipping from which alone a military fleet naturally and healthfully springs, and on which it securely rests.

The principal conditions affecting the sea power of nations may be enumerated as follows: I. Geographical Position. II. Physical Conformation, including, as connected therewith, natural productions and climate. III. Extent of Territory. IV. Number of Population. V. Character of the People. VI. Character of the Government, including therein the national institutions.

1. Geographical Position. It may be pointed out, in the first place, that if a nation be so situated that it is neither forced to defend itself by land nor induced to seek extension of its territory by way of the land, it has, by the very unity of its aim directed upon the sea, an advantage as compared with a people one of whose boundaries is continental. This has been a great advantage to England over both France and Holland as a sea power. The strength of the latter was early exhausted by the necessity of keeping up a large army and carrying on expensive wars to preserve her independence; while the policy of France was constantly diverted, sometimes wisely and sometimes most foolishly, from the sea to projects of continental extension. These military efforts expended wealth; whereas a wiser and consistent use of her geographical position would have added to it.

The geographical position may be such as of itself to promote a concentration, or to necessitate a dispersion, of the naval forces. Here again the British Islands have an advantage over France. The position of the latter, touching the Mediterranean as well as the ocean, while it has its advantages, is on the whole a source of military weakness at sea. The eastern and western French fleets have only been able to unite after passing through the Straits of Gibraltar, in attempting which they have often risked and sometimes suffered loss. The position of the United States upon the two oceans would be either a source of great weakness or a cause of enormous expense, had it a large sea commerce on both coasts.

England, by her immense colonial empire, has sacrificed much of this advantage of concentration of force around her own shores; but the sacrifice was wisely made, for the gain was greater than the loss, as the event proved. With the growth of her colonial system her war fleets also grew, but her merchant shipping and wealth grew yet faster. Still, in the wars of the American Revolution, and of the French Republic and Empire, to use the strong expression of a French author, "England, despite the immense development of her navy, seemed ever, in the midst of riches, to feel all the embarrassment of poverty." The might of England was sufficient to keep alive the heart and the members; whereas the equally extensive colonial empire of Spain, through her maritime weakness, but offered so many points for insult and injury.

The geographical position of a country may not only favor the concentration of its forces, but give the further strategic advantage of a central position and a good base for hostile operations against its probable enemies. This again is the case with England; on the one hand she faces Holland and the northern powers, on the other France and the Atlantic. When threatened with a coalition between France and the naval powers of the North Sea and the Baltic, as she at times was, her fleets in the Downs and in the Channel, and even that off Brest, occupied interior positions, and thus were readily able to interpose their united force against either one of the enemies which should seek to pass through the Channel to effect a junction with its ally. On either side, also, Nature gave her better ports and a safer coast to approach. Formerly this was a very serious element in the passage through the Channel; but of late, steam and the improvement of her harbors have lessened the disadvantage under which France once labored. In the days of sailing-ships, the English fleet operated against Brest making its base at Torbay and Plymouth. The plan was simply this: in easterly or moderate weather the blockading fleet kept its position without difficulty; but in westerly gales, when too severe, they bore up for English ports, knowing that the French fleet could not get out till the wind shifted, which equally served to bring them back to their station.

The advantage of geographical nearness to an enemy, or to the object of attack, is nowhere more apparent than in that form of warfare which has lately received the name of commerce-destroying, which the French call *guerre de course*. This operation of war, being directed against peaceful merchant vessels which are usually defenceless, calls for ships of small military force. Such ships, having little power to defend themselves, need a refuge or point of support near at hand; which will be found either in certain parts of the sea controlled by the fighting ships of their country, or in friendly harbors. The latter give the strongest support, because they are always in the same place, and the approaches to them are more familiar to the commerce-destroyer than to his enemy. The nearness of France to England has thus greatly facilitated her *guerre de course* directed against the latter. Having ports on the North Sea, on the Channel, and on the Atlantic, her cruisers started from points near the focus of English trade, both coming and going. The distance of these ports from each other, disadvantageous for regular military combinations, is an advantage for this irregular secondary operation; for the essence of the one is concentration of effort, whereas for commerce-destroying diffusion of effort is the rule. Commerce-destroyers scatter, that they may see and seize more prey. These truths receive illustration from the history of the great French privateers, whose bases and scenes of action were largely on the Channel and North Sea, or else were found in distant colonial regions, where islands like Guadaloupe and Martinique afforded similar near refuge. The necessity of renewing coal makes the cruiser of the present day even more dependent than of old on his port. Public opinion in the United States has great faith in war directed against an enemy's commerce; but it must be remembered that the Republic has no ports very near the great centres of trade abroad. Her geographical position is therefore singularly disadvantageous for carrying on successful commerce-destroying, unless she find bases in the ports of an ally.

If, in addition to facility for offence, Nature has so placed a country that it has easy access to the high sea itself, while at the same time it controls one of the great thoroughfares of the world's traffic, it is evident that the strategic value of its position is very high. Such again is, and to a greater degree was, the position of England. The trade of Holland, Sweden, Russia, Denmark, and that which went up the great rivers to the interior of Germany, had to pass through the Channel close by her doors; for sailing-ships hugged the English coast. This northern trade had, moreover, a peculiar bearing upon sea power; for naval stores, as they are commonly called, were mainly drawn from the Baltic countries.

But for the loss of Gibraltar, the position of Spain would have been closely analogous to that of England. Looking at once upon the Atlantic and the Mediterranean, with Cadiz on the one side and Cartagena on the other, the trade to the Levant must have passed under her hands, and that round the Cape of Good Hope not far from her doors. But Gibraltar not only deprived her of the control of the Straits, it also imposed an obstacle to the easy junction of the two divisions of her fleet.

At the present day, looking only at the geographical position of Italy, and not at the other conditions affecting her sea power, it would seem that with her extensive sea-coast and good ports she is very well placed for exerting a decisive influence on the trade route to the Levant and by the Isthmus of Suez. This is true in a degree, and would be much more so did Italy now hold all the islands naturally Italian; but with Malta in the hands of England, and Corsica in those of France, the advantages of her geographical position are largely neutralized. From race affinities and situation those two islands are as legitimately objects of desire to Italy as Gibraltar is to Spain. If the Adriatic were a great highway of commerce, Italy's position would be still more influential. These defects in her geographical completeness, combined with other causes injurious to a full and secure development of sea power, make it more than doubtful whether Italy can for some time be in the front rank among the sea nations.

As the aim here is not an exhaustive discussion, but merely an attempt to show, by illustration, how vitally the situation of a country may affect its career upon the sea, this division of the subject may be dismissed for the present; the more so as instances which will further bring out its importance will continually recur in the historical treatment. Two remarks, however, are here appropriate.

Circumstances have caused the Mediterranean Sea to play a greater part in the history of the world, both in a commercial and a military point of view, than any other sheet of water of the same size. Nation after nation has striven to control it, and the strife still goes on. Therefore a study of the conditions upon which preponderance in its waters has rested, and now rests, and of the relative military values of different points upon its coasts, will be more instructive than the same amount of effort expended in another field. Furthermore, it has at the present time a very marked analogy in many respects to the Caribbean Sea,—an analogy which will be still closer if a Panama canal-route ever be completed. A study of the strategic conditions of the Mediterranean, which have received ample illustration, will be an excellent prelude to a similar study of the Caribbean, which has comparatively little history.

The second remark bears upon the geographical position of the United States relatively to a Central-American canal. If one be made, and fulfil the hopes of its builders, the Caribbean will be changed from a terminus, and place of local traffic, or at best a broken and imperfect line of travel, as it now is, into one of the great highways of the world. Along this path a great commerce will travel, bringing the interests of the other great nations, the European nations, close along our shores, as they have never been before. With this it will not be so easy as heretofore to stand aloof from international complications. The position of the United States with reference to this route will resemble that of England to the Channel, and of the Mediterranean countries to the Suez route. As regards influence and control over it, depending upon geographical position, it is of course plain that the centre of the national power, the permanent base,[1] is much nearer than that of other great nations. The positions now or hereafter occupied by them on island or mainland, however strong, will be but outposts of their power; while in all the raw materials of military strength no nation is superior to the United States. She is, however, weak in a confessed unpreparedness for war; and her geographical nearness to the point of contention loses some of its value by the character of the Gulf coast, which is deficient in ports combining security from an enemy with facility for repairing war-ships of the first class, without which ships no country can pretend to control any part of the sea. In case of a contest for supremacy in the Caribbean, it seems evident from the depth of the South Pass of the Mississippi, the nearness of New Orleans, and the advantages of the Mis-

1. By a base of permanent operations "is understood a country whence come all the resources, where are united the great lines of communication by land and water, where are the arsenals and armed posts."

sissippi Valley for water transit, that the main effort of the country must pour down that valley, and its permanent base of operations be found there. The defence of the entrance to the Mississippi, however, presents peculiar difficulties; while the only two rival ports, Key West and Pensacola, have too little depth of water, and are much less advantageously placed with reference to the resources of the country. To get the full benefit of superior geographical position, these defects must be overcome. Furthermore, as her distance from the Isthmus, though relatively less, is still considerable, the United States will have to obtain in the Caribbean stations fit for contingent, or secondary, bases of operations; which by their natural advantages, susceptibility of defence, and nearness to the central strategic issue, will enable her fleets to remain as near the scene as any opponent. With ingress and egress from the Mississippi sufficiently protected, with such outposts in her hands, and with the communications between them and the home base secured, in short, with proper military preparation, for which she has all necessary means, the preponderance of the United States on this field follows, from her geographical position and her power, with mathematical certainty.

2. *Physical Conformation.* The peculiar features of the Gulf coast, just alluded to, come properly under the head of Physical Conformation of a country, which is placed second for discussion among the conditions which affect the development of sea power.

The seaboard of a country is one of its frontiers; and the easier the access offered by the frontier to the region beyond, in this case the sea, the greater will be the tendency of a people toward intercourse with the rest of the world by it. If a country be imagined having a long seaboard, but entirely without a harbor, such a country can have no sea trade of its own, no shipping, no navy. This was practically the case with Belgium when it was a Spanish and an Austrian province. The Dutch, in 1648, as a condition of peace after a successful war, exacted that the Scheldt should be closed to sea commerce. This closed the harbor of Antwerp and transferred the sea trade of Belgium to Holland. The Spanish Netherlands ceased to be a sea power.

Numerous and deep harbors are a source of strength and wealth, and doubly so if they are the outlets of navigable streams, which facilitate the concentration in them of a country's internal trade; but by their very accessibility they become a source of weakness in war, if not properly defended. The Dutch in 1667 found little difficulty in ascending the Thames and burning a large fraction of the English navy within sight of London; whereas a few years later the combined fleets of England and France, when attempting a landing in Holland, were foiled by the difficulties of the coast as much as by the valor of the Dutch fleet. In 1778 the harbor of New York, and with it undisputed control of the Hudson River, would have been lost to the English, who were caught at disadvantage, but for the hesitancy of the French admiral. With that control, New England would have been restored to close and safe communication with New York, New Jersey, and Pennsylvania; and this blow, following so closely on Burgoyne's disaster of the year before, would probably have led the English to make an earlier peace. The Mississippi is a mighty source of wealth and strength to the United States; but the feeble defences of its mouth and the number of its subsidiary streams penetrating the country made it a weakness and source of disaster to the Southern Confederacy. And lastly, in 1814, the occupation of the Chesapeake and the destruction of Washington gave a sharp lesson of the dangers incurred through the noblest water-ways, if their approaches be undefended; a lesson recent enough to be easily recalled, but which, from the present appearance of the coast defences, seems to be yet more easily forgotten. Nor should it be thought that conditions have changed; circumstances and details of offence and defence have been modified, in these days as before, but the great conditions remain the same.

Before and during the great Napoleonic wars, France had no port for ships-of-the-line east of Brest. How great the advantage to England, which in the same stretch has two great arsenals, at Plymouth and at Portsmouth, besides other harbors of refuge and supply. This defect of conformation has since been remedied by the works at Cherbourg.

Besides the contour of the coast, involving easy access to the sea, there are other physical conditions which lead people to the sea or turn them from it. Although France was deficient in military ports on the Channel, she had both there and on the ocean, as well as in the Mediterranean, excellent harbors, favorably situated for trade abroad, and at the outlet of large rivers, which would foster internal traffic. But when Richelieu had put an end to civil war, Frenchmen did not take to the sea with the eagerness and success of the English and Dutch. A principal reason for this has been plausibly found in the physical conditions which have made France a pleasant land, with a delightful climate, producing within itself more than its people needed. England, on the other hand, received from Nature but little, and, until her manufactures were developed, had little to export. Their many wants, combined with their restless activity and other conditions that favored maritime enterprise, led her people abroad; and they there found lands more pleasant and richer than their own. Their needs and genius made them merchants and colonists, then manufacturers and producers; and between products and colonies shipping is the inevitable link. So their sea power grew. But if England was drawn to the sea, Holland was driven to it; without the sea England languished, but Holland died. In the height of her greatness, when she was one of the chief factors in European politics, a competent native authority estimated that the soil of Holland could not support more than one eighth of her inhabitants. The manufactures of the country were then numerous and important, but they had been much later in their growth than the shipping interest. The poverty of the soil and the exposed nature of the coast drove the Dutch first to fishing. Then the discovery of the process of curing the fish gave them material for export as well as home consumption, and so laid the corner-stone of their wealth. Thus they had become traders at the time that the Italian republics, under the pressure of Turkish power and the discovery of the passage round the Cape of Good Hope, were beginning to decline, and they fell heirs to the great Italian trade of the Levant. Further favored by their geographical position, intermediate between the Baltic, France, and the Mediterranean, and at the mouth of the German rivers, they quickly absorbed nearly all the carrying-trade of Europe. The wheat and naval stores of the Baltic, the trade of Spain with her colonies in the New World, the wines of France, and the French coasting-trade were, little more than two hundred years ago, transported in Dutch shipping. Much of the carrying-trade of England, even, was then done in Dutch bottoms. It will not be pretended that all this prosperity proceeded only from the poverty of Holland's natural resources. Something does not grow from nothing. What is true, is, that by the necessitous condition of her people they were driven to the sea, and were, from their mastery of the shipping business and the size of their fleets, in a position to profit by the sudden expansion of commerce and the spirit of exploration which followed on the discovery of America and of the passage round the Cape. Other causes concurred, but their whole prosperity stood on the sea power to which their poverty gave birth. Their food, their clothing, the raw material for their manufactures, the very timber and hemp with which they built and rigged their ships (and they built nearly as many as all Europe besides), were imported; and when a disastrous war with England in 1653 and 1654 had lasted eighteen months, and their shipping business was stopped, it is said "the sources of revenue which had always maintained the riches of the State, such as fisheries and commerce, were almost dry. Workshops were closed, work was suspended. The Zuyder Zee became a forest of masts; the country was full of beggars; grass grew in the streets, and in Amsterdam fifteen hundred houses were untenanted." A humiliating peace alone saved them from ruin.

This sorrowful result shows the weakness of a country depending wholly upon sources external to itself for the part it is playing in the world. With large deductions, owing to differences of conditions which need not here be spoken of, the case of Holland then has strong points of resemblance to that of Great Britain now; and they are true prophets, though they seem to be having small honor in their own country, who warn her that the continuance of her prosperity at home depends primarily upon maintaining her power abroad. Men may be discontented at the lack of political privilege; they will be yet more uneasy if they come to lack bread. It is of more interest to Americans to note that the result to France, regarded as a power of the sea, caused by the extent, delightfulness, and richness of the land, has been reproduced in the United States. In the beginning, their forefathers held a narrow strip of land upon the sea, fertile in parts though little developed, abounding in harbors and near rich fishing-grounds. These physical conditions combined with an inborn love of the sea, the pulse of that English blood which still beat in their veins, to keep alive all those tendencies and pursuits upon which a healthy sea power depends. Almost every one of the original colonies was on the sea or on one of its great tributaries. All export and import tended toward one coast. Interest in the sea and an intelligent appreciation of the part it played in the public welfare were easily and widely spread; and a motive more influential than care for the public interest was also active, for the abundance of ship-building materials and a relative fewness of other investments made shipping a profitable private interest. How changed the present condition is, all know. The centre of power is no longer on the seaboard. Books and newspapers vie with one another in describing the wonderful growth, and the still undeveloped riches, of the interior. Capital there finds its best investments, labor its largest opportunities. The frontiers are neglected and politically weak; the Gulf and Pacific coasts actually so, the Atlantic coast relatively to the central Mississippi Valley. When the day comes that shipping again pays, when the three sea frontiers find that they are not only militarily weak, but poorer for lack of national shipping, their united efforts may avail to lay again the foundations of our sea power. Till then, those who follow the limitations which lack of sea power placed upon the career of France may mourn that their own country is being led, by a like redundancy of home wealth, into the same neglect of that great instrument.

Among modifying physical conditions may be noted a form like that of Italy,—a long peninsula, with a central range of mountains dividing it into two narrow strips, along which the roads connecting the different ports necessarily run. Only an absolute control of the sea can wholly secure such communications, since it is impossible to know at what point an enemy coming from beyond the visible horizon may strike; but still, with an adequate naval force centrally posted, there will be good hope of attacking his fleet, which is at once his base and line of communications, before serious damage has been done. The long, narrow peninsula of Florida, with Key West at its extremity, though flat and thinly populated, presents at first sight conditions like those of Italy. The resemblance may be only superficial, but it seems probable that if the chief scene of a naval war were the Gulf of Mexico, the communications by land to the end of the peninsula might be a matter of consequence, and open to attack.

When the sea not only borders, or surrounds, but also separates a country into two or more parts, the control of it becomes not only desirable, but vitally necessary. Such a physical condition either gives birth and strength to sea power, or makes the country powerless. Such is the condition of the present kingdom of Italy, with its islands of Sardinia and Sicily; and hence in its youth and still existing financial weakness it is seen to put forth such vigorous and intelligent efforts to create a military navy. It has even been argued that, with a navy decidedly superior to her enemy's, Italy could better base her power upon her islands than upon her mainland; for the insecurity of the lines of communication in the peninsula, already pointed out, would most seriously embarrass an invading army surrounded by a hostile people and threatened from the sea.

The Irish Sea, separating the British Islands, rather resembles an estuary than an actual division; but history has shown the danger from it to the United Kingdom. In the days of Louis XIV., when the French navy nearly equalled the combined English and Dutch, the gravest complications existed in Ireland, which passed almost wholly under the control of the natives and the French. Nevertheless, the Irish Sea was rather a danger to the English—a weak point in their communications—than an advantage to the French. The latter did not venture their ships-of-the-line in its narrow waters, and expeditions intending to land were directed upon the ocean ports in the south and west. At the supreme moment the great French fleet was sent upon the south coast of England, where it decisively defeated the allies, and at the same time twenty-five frigates were sent to St. George's Channel, against the English communications. In the midst of a hostile people, the English army in Ireland was seriously imperilled, but was saved by the battle of the Boyne and the flight of James II. This movement against the enemy's communications was strictly strategic, and would be just as dangerous to England now as in 1690.

Spain, in the same century, afforded an impressive lesson of the weakness caused by such separation when the parts are not knit together by a strong sea power. She then still retained, as remnants of her past greatness, the Netherlands (now Belgium), Sicily, and other Italian possessions, not to speak of her vast colonies in the New World. Yet so low had the Spanish sea power fallen, that a well-informed and sober-minded Hollander of the day could claim that "in Spain all the coast is navigated by a few Dutch ships; and since the peace of 1648 their ships and seamen are so few that they have publicly begun to hire our ships to sail to the Indies, whereas they were formerly careful to exclude all foreigners from there. . . . It is manifest," he goes on, "that the West Indies, being as the stomach to Spain (for from it nearly all the revenue is drawn), must be joined to the Spanish head by a sea force; and that Naples and the Netherlands, being like two arms, they cannot lay out their strength for Spain, nor receive anything thence but by shipping,—all which may easily be done by our shipping in peace, and by it obstructed in war." Half a century before, Sully, the great minister of Henry IV., had characterized Spain "as one of those States whose legs and arms are strong and powerful, but the heart infinitely weak and feeble." Since his day the Spanish navy had suffered not only disaster, but annihilation; not only humiliation, but degradation. The consequences briefly were that shipping was destroyed; manufactures perished with it. The government depended for its support, not upon a widespread healthy commerce and industry that could survive many a staggering blow, but upon a narrow stream of silver trickling through a few treasure-ships from America, easily and frequently intercepted by an enemy's cruisers. The loss of half a dozen galleons more than once paralyzed its movements for a year. While the war in the Netherlands lasted, the Dutch control of the sea forced Spain to send her troops by a long and costly journey overland instead of by sea; and the same cause reduced her to such straits for necessaries that, by a mutual arrangement which seems very odd to modern ideas, her wants were supplied by Dutch ships, which thus maintained the enemies of their country, but received in return specie which was welcome in the Amsterdam exchange. In America, the Spanish protected themselves as best they might behind masonry, unaided from home; while in the Mediterranean they escaped insult and injury mainly through the indifference of the Dutch, for the French and English had not yet begun to contend for mastery there. In the course of history the Netherlands, Naples, Sicily, Minorca, Havana, Manila, and Jamaica were wrenched away, at one time or another, from this empire without a shipping. In short, while Spain's maritime impotence may have been primarily a symptom of her general decay, it became a marked factor in precipitating her into the abyss from which she has not yet wholly emerged.

Except Alaska, the United States has no outlying possession,—no foot of ground inaccessible by land. Its contour is such as to present few points specially weak from their saliency, and all important parts of the frontiers can be readily attained,—cheaply by water, rapidly by rail. The weakest frontier, the Pacific, is far removed from the most dangerous of possible enemies. The internal resources are boundless as compared with present needs; we can live off ourselves indefinitely in "our little corner," to use the expression of a French officer to the author. Yet should that little corner be invaded by a new commercial route through the Isthmus, the United States in her turn may have the rude awakening of those who have abandoned their share in the common birthright of all people, the sea.

3. Extent of Territory. The last of the conditions affecting the development of a nation as a sea power, and touching the country itself as distinguished from the people who dwell there, is Extent of Territory. This may be dismissed with comparatively few words.

As regards the development of sea power, it is not the total number of square miles which a country contains, but the length of its coast-line and the character of its harbors that are to be considered. As to these it is to be said that, the geographical and physical conditions being the same, extent of sea-coast is a source of strength or weakness according as the population is large or small. A country is in this like a fortress; the garrison must be proportioned to the *enceinte*. A recent familiar instance is found in the American War of Secession. Had the South had a people as numerous as it was warlike, and a navy commensurate to its other resources as a sea power, the great extent of its sea-coast and its numerous inlets would have been elements of great strength. The people of the United States and the Government of that day justly prided themselves on the effectiveness of the blockade of the whole Southern coast. It was a great feat, a very great feat; but it would have been an impossible feat had the Southerners been more numerous, and a nation of seamen. What was there shown was not, as has been said, how such a blockade can be maintained, but that such a blockade is possible in the face of a population not only unused to the sea, but also scanty in numbers. Those who recall how the blockade was maintained, and the class of ships that blockaded during great part of the war, know that the plan, correct under the circumstances, could not have been carried out in the face of a real navy. Scattered unsupported along the coast, the United States ships kept their places, singly or in small detachments, in face of an extensive network of inland water communications which favored secret concentration of the enemy. Behind the first line of water communications were long estuaries, and here and there strong fortresses, upon either of which the enemy's ships could always fall back to elude pursuit or to receive protection. Had there been a Southern navy to profit by such advantages, or by the scattered condition of the United States ships, the latter could not have been distributed as they were; and being forced to concentrate for mutual support, many small but useful approaches would have been left open to commerce. But as the Southern coast, from its extent and many inlets, might have been a source of strength, so, from those very characteristics, it became a fruitful source of injury. The great story of the opening of the Mississippi is but the most striking illustration of an action that was going on incessantly all over the South. At every breach of the sea frontier, war-ships were entering. The streams that had carried the wealth and supported the trade of the seceding States turned against them, and admitted their enemies to their hearts. Dismay, insecurity, paralysis, prevailed in regions that might, under happier auspices, have kept a nation alive through the most exhausting war. Never did sea power play a greater or a more decisive part than in the contest which determined that the course of the world's history would be modified by the existence of one great nation, instead of several rival States, in the North American continent. But while just pride is felt in the well-earned glory of those days, and the greatness of the results due to naval preponderance is admitted, Americans who understand the facts should never fail to remind the over-confidence of their countrymen that the South not only had no navy, not only was not a seafaring people, but that also its population was not proportioned to the extent of the sea-coast which it had to defend.

4. Number of Population. After the consideration of the natural conditions of a country should follow an examination of the characteristics of its population as affecting the development of sea power; and first among these will be taken, because of its relations to the extent of the territory, which has just been discussed, the number of the people who live in it. It has been said that in respect of dimensions it is not merely the number of square miles, but the extent and character of the sea-coast that is to be considered with reference to sea power;

and so, in point of population, it is not only the grand total, but the number following the sea, or at least readily available for employment on ship-board and for the creation of naval material, that must be counted.

For example, formerly and up to the end of the great wars following the French Revolution, the population of France was much greater than that of England; but in respect of sea power in general, peaceful commerce as well as military efficiency, France was much inferior to England. In the matter of military efficiency this fact is the more remarkable because at times, in point of military preparation at the outbreak of war, France had the advantage; but she was not able to keep it. Thus in 1778, when war broke out, France, through her maritime inscription, was able to man at once fifty ships-of-the-line. England, on the contrary, by reason of the dispersal over the globe of that very shipping on which her naval strength so securely rested, had much trouble in manning forty at home; but in 1782 she had one hundred and twenty in commission or ready for commission, while France had never been able to exceed seventy-one. Again, as late as 1840, when the two nations were on the verge of war in the Levant, a most accomplished French officer of the day, while extolling the high state of efficiency of the French fleet and the eminent qualities of its admiral, and expressing confidence in the results of an encounter with an equal enemy, goes on to say: "Behind the squadron of twenty-one ships-of-the-line which we could then assemble, there was no reserve; not another ship could have been commissioned within six months." And this was due not only to lack of ships and of proper equipments, though both were wanting. "Our maritime inscription," he continues, "was so exhausted by what we had done [in manning twenty-one ships], that the permanent levy established in all quarters did not supply reliefs for the men, who were already more than three years on cruise."

A contrast such as this shows a difference in what is called staying power, or reserve force, which is even greater than appears on the surface; for a great shipping afloat necessarily employs, besides the crews, a large number of people engaged in the various handicrafts which facilitate the making and repairing of naval material, or following other callings more or less closely connected with the water and with craft of all kinds. Such kindred callings give an undoubted aptitude for the sea from the outset. There is an anecdote showing curious insight into this matter on the part of one of England's distinguished seamen, Sir Edward Pellew. When the war broke out in 1793, the usual scarceness of seamen was met. Eager to get to sea and unable to fill his complement otherwise than with landsmen, he instructed his officers to seek for Cornish miners; reasoning from the conditions and dangers of their calling, of which he had personal knowledge, that they would quickly fit into the demands of sea life. The result showed his sagacity, for, thus escaping an otherwise unavoidable delay, he was fortunate enough to capture the first frigate taken in the war in single combat; and what is especially instructive is, that although but a few weeks in commission, while his opponent had been over a year, the losses, heavy on both sides, were nearly equal.

It may be urged that such reserve strength has now nearly lost the importance it once had, because modern ships and weapons take so long to make, and because modern States aim at developing the whole power of their armed force, on the outbreak of war, with such rapidity as to strike a disabling blow before the enemy can organize an equal effort. To use a familiar phrase, there will not be time for the whole resistance of the national fabric to come into play; the blow will fall on the organized military fleet, and if that yield, the solidity of the rest of the structure will avail nothing. To a certain extent this is true; but then it has always been true, though to a less extent formerly than now. Granted the meeting of two fleets which represent practically the whole present strength of their two nations, if one of them be destroyed, while the other remains fit for action, there will be much less hope now than formerly that the vanquished can restore his navy for that war; and the result will be disastrous just in proportion to the dependence of the nation upon her sea power. A Trafalgar would have been a much more fatal blow to England than it was to France, had the English fleet then represented, as the allied fleet did, the bulk of the nation's power. Trafalgar in such a case would have been to England what Austerlitz was to Austria, and Jena to Prussia; an empire would have been laid prostrate by the destruction or disorganization of its military forces, which, it is said, were the favorite objective of Napoleon.

But does the consideration of such exceptional disasters in the past justify the putting a low value upon that reserve strength, based upon the number of inhabitants fitted for a certain kind of military life, which is here being considered? The blows just mentioned were dealt by men of exceptional genius, at the head of armed bodies of exceptional training, *esprit-de-corps,* and prestige, and were, besides, inflicted upon opponents more or less demoralized by conscious inferiority and previous defeat. Austerlitz had been closely preceded by Ulm, where thirty thousand Austrians laid down their arms without a battle; and the history of the previous years had been one long record of Austrian reverse and French success. Trafalgar followed closely upon a cruise, justly called a campaign, of almost constant failure; and farther back, but still recent, were the memories of St. Vincent for the Spaniards, and of the Nile for the French, in the allied fleet. Except the case of Jena, these crushing overthrows were not single disasters, but final blows; and in the Jena campaign there was a disparity in numbers, equipment, and general preparation for war, which makes it less applicable in considering what may result from a single victory.

England is at the present time the greatest maritime nation in the world; in steam and iron she has kept the superiority she had in the days of sail and wood. France and England are the two powers that have the largest military navies; and it is so far an open question which of the two is the more powerful, that they may be regarded as practically of equal strength in material for a sea war. In the case of a collision can there be assumed such a difference of *personnel,* or of preparation, as to make it probable that a decisive inequality will result from one battle or one campaign? If not, the reserve strength will begin to tell; organized reserve first, then reserve of seafaring population, reserve of mechanical skill, reserve of wealth. It seems to have been somewhat forgotten that England's leadership in mechanical arts gives her a reserve of mechanics, who can easily familiarize themselves with the appliances of modern iron-clads; and as her commerce and industries feel the burden of the war, the surplus of seamen and mechanics will go to the armed shipping.

The whole question of the value of a reserve, developed or undeveloped, amounts now to this: Have modern conditions of warfare made it probable that, of two nearly equal adversaries, one will be so prostrated in a single campaign that a decisive result will be reached in that time? Sea warfare has given no answer. The crushing successes of Prussia against Austria, and of Germany against France, appear to have been those of a stronger over a much weaker nation, whether the weakness were due to natural causes, or to official incompetency. How would a delay like that of Plevna have affected the fortune of war, had Turkey had any reserve of national power upon which to call?

If time be, as is everywhere admitted, a supreme factor in war, it behooves countries whose genius is essentially not military, whose people, like all free people, object to pay for large military establishments, to see to it that they are at least strong enough to gain the time necessary to turn the spirit and capacity of their subjects into the new activities which war calls for. If the existing force by land or sea is strong enough so to hold out, even though at a disadvantage, the country may rely upon its natural resources and strength coming into play for whatever they are worth,—its numbers, its wealth, its capacities of every kind. If, on the other hand, what force it has can be overthrown and crushed quickly, the most magnificent possibilities of natural power will not save it from humiliating conditions, nor, if its foe be wise, from guarantees which will postpone revenge to a distant future. The story is constantly repeated on the smaller fields of war: "If so-and-so can hold out a little longer, this can be saved or that can be done;" as in sickness it is often said: "If the patient can only hold out so long, the strength of his constitution may pull him through."

England to some extent is now such a country. Holland was such a country; she would not pay, and if she escaped, it was but by the skin of her teeth. "Never in time of peace and from fear of a rupture," wrote their great statesman, De Witt, "will they take resolutions strong enough to lead them to pecuniary sacrifices beforehand. The character of the Dutch is such that, unless danger stares them in the face, they are indisposed to lay out money for their own defence. I have to do with a people who, liberal to profusion where they ought to economize, are often sparing to avarice where they ought to spend."

That our own country is open to the same reproach, is patent to all the world. The United States has not that shield of defensive power behind which time can be gained to develop its reserve of strength. As for a seafaring population adequate to her possible needs, where is it? Such a resource, proportionate to her coast-line and population, is to be found only in a national merchant shipping and its related industries, which at present scarcely exist. It will matter little whether the crews of such ships are native or foreign born, provided they are attached to the flag, and her power at sea is sufficient to enable the most of them to get back in case of war. When foreigners by thousands are admitted to the ballot, it is of little moment that they are given fighting-room on board ship.

Though the treatment of the subject has been somewhat discursive, it may be admitted that a great population following callings related to the sea is, now as formerly, a great element of sea power; that the United States is deficient in that element; and that its foundations can be laid only in a large commerce under her own flag.

5. National Character. The effect of national character and aptitudes upon the development of sea power will next be considered.

If sea power be really based upon a peaceful and extensive commerce, aptitude for commercial pursuits must be a distinguishing feature of the nations that have at one time or another been great upon the sea. History almost without exception affirms that this is true. Save the Romans, there is no marked instance to the contrary.

All men seek gain and, more or less, love money; but the way in which gain is sought will have a marked effect upon the commercial fortunes and the history of the people inhabiting a country.

If history may be believed, the way in which the Spaniards and their kindred nation, the Portuguese, sought wealth, not only brought a blot upon the national character, but was also fatal to the growth of a healthy commerce; and so to the industries upon which commerce lives, and ultimately to that national wealth which was sought by mistaken paths. The desire for gain rose in them to fierce avarice; so they sought in the new-found worlds which gave such an impetus to the commercial and maritime development of the countries of Europe, not new fields of industry, not even the healthy excitement of exploration and adventure, but gold and silver. They had many great qualities; they were bold, enterprising, temperate, patient of suffering, enthusiastic, and gifted with intense national feeling. When to these qualities are added the advantages of Spain's position and well-situated ports, the fact that she was first to occupy large and rich portions of the new worlds and long remained without a competitor, and that for a hundred years after the discovery of America she was the leading State in Europe, she might have been expected to take the foremost place among the sea powers. Exactly the contrary was the result, as all know. Since the battle of Lepanto in 1571, though engaged in many wars, no sea victory of any consequence shines on the pages of Spanish history; and the decay of her commerce sufficiently accounts for the painful and sometimes ludicrous inaptness shown on the decks of her ships of war. Doubtless such a result is not to be attributed to one cause only. Doubtless the government of Spain was in many ways such as to cramp and blight a free and healthy development of private enterprise; but the character of a great people breaks through or shapes the character of its government, and it can hardly be doubted that had the bent of the people been toward trade, the action of government would have been drawn into the same current. The great field of the colonies, also, was remote from the centre of that despotism which blighted the growth of old Spain. As it was, thousands of Spaniards, of the working as well as the upper classes, left Spain; and the occupations in which they engaged abroad sent home little but specie, or merchandise of small bulk, requiring but small tonnage. The mother-country herself produced little but wool, fruit, and iron; her manufactures were naught; her industries suffered; her population steadily decreased. Both she and her colonies depended upon the Dutch for so many of the necessaries of life, that the products of their scanty industries could not suffice to pay for them. "So that Holland merchants," writes a contemporary, "who carry money to most parts of the world to buy commodities, must out of this single country of Europe carry home money, which they receive in payment of their goods." Thus their eagerly sought emblem of wealth passed quickly from their hands. It has already been pointed out how weak, from a military point of view, Spain was from this decay of her shipping. Her wealth being in small bulk on a few ships, following more or less regular routes, was easily seized by an enemy, and the sinews of war paralyzed; whereas the wealth of England and Holland, scattered over thousands of ships in all parts of the world, received many bitter blows in many exhausting wars, without checking a growth which, though painful, was steady. The fortunes of Portugal, united to Spain during a most critical period of her history, followed the same downward path; although foremost in the beginning of the race for development by sea, she fell utterly behind. "The mines of Brazil were the ruin of Portugal, as those of Mexico and Peru had been of Spain; all manufactures fell into insane contempt; ere long the English supplied the Portuguese not only with clothes, but with all merchandise, all commodities, even to salt-fish and grain. After their gold, the Portuguese abandoned their very soil; the vineyards of Oporto were finally bought by the English with Brazilian gold, which had only passed through Portugal to be spread throughout England." We are assured that in fifty years, five hundred millions of dollars were extracted from "the mines of Brazil, and that at the end of the time Portugal had but twenty-five millions in specie,"—a striking example of the difference between real and fictitious wealth.

The English and Dutch were no less desirous of gain than the southern nations. Each in turn has been called "a nation of shopkeepers;" but the jeer, in so far as it is just, is to the credit of their wisdom and uprightness. They were no less bold, no less enterprising, no less patient. Indeed, they were more patient, in that they sought riches not by the sword but by labor, which is the reproach meant to be implied by the epithet; for thus they took the longest, instead of what seemed the shortest, road to wealth. But these two peoples, radically of the same race, had other qualities, no less important than those just named, which combined with their surroundings to favor their development by sea. They were by nature business-men, traders, producers, negotiators. Therefore both in their native country and abroad, whether settled in the ports of civilized nations, or of barbarous eastern rulers, or in colonies of their own foundation, they everywhere strove to draw out all the resources of the land, to develop and increase them. The quick instinct of the born trader, shopkeeper if you will, sought continually new articles to exchange; and this search, combined with the industrious character evolved through generations of labor, made them necessarily producers. At home they became great as manufacturers; abroad, where they controlled, the land grew richer continually, products multiplied, and the necessary exchange between home and the settlements called for more ships. Their shipping therefore increased with these demands of trade, and nations with less aptitude for maritime enterprise, even France herself, great as she has been, called for their products and for the service of their ships. Thus in many ways they advanced to power at sea. This natural tendency and growth were indeed modified and seriously checked at times by the interference of other governments, jealous of a prosperity which their own people could invade only by the aid of artificial support,—a support which will be considered under the head of governmental action as affecting sea power.

The tendency to trade, involving of necessity the production of something to trade with, is the national characteristic most important to the development of sea power. Granting it and a good seaboard, it is not likely that the dangers of the sea, or any aversion to it, will deter a people from seeking wealth by the paths of ocean commerce. Where wealth is sought by other means, it may be found; but it will not necessarily lead to sea power. Take France. France has a fine country, an industrious people, an admirable position. The French navy has known periods of great glory, and in its lowest estate has never dishonored the military reputation so dear to the nation. Yet as a maritime State, securely resting upon a broad basis of sea commerce, France, as compared with other historical sea-peoples, has never held more than a respectable position. The chief reason for this, so far as national character goes, is the way in which wealth is sought. As Spain and Portugal sought it by digging gold out of the ground, the temper of the French people leads them to seek it by thrift, economy, hoarding. It is said to be harder to keep than to make a fortune. Possibly; but the adventurous temper, which risks what it has to gain more, has much in common with the adventurous spirit that conquers worlds for commerce. The tendency to save and put aside, to venture timidly and on a small scale, may lead to a general diffusion of wealth on a like small scale, but not to the risks and development of external trade and shipping interests. To illustrate,—and the incident is given only for what it is worth,—a French officer, speaking to the author abut the Panama Canal, said: "I have two shares in it. In France we don't do as you, where a few people take a great many shares each. With us a large number of people take one share or a very few. When these were in the market my wife said to me, 'You take two shares, one for you and one for me.' " As regards the stability of a man's personal fortunes this kind of prudence is doubtless wise; but when excessive prudence or financial timidity becomes a national trait, it must tend to hamper the expansion of commerce and of the nation's shipping. The same caution in money matters, appearing in another relation of life, has checked the production of children, and keeps the population of France nearly stationary.

The noble classes of Europe inherited from the Middle Ages

a supercilious contempt for peaceful trade, which has exercised a modifying influence upon its growth, according to the national character of different countries. The pride of the Spaniards fell easily in with this spirit of contempt, and cooperated with that disastrous unwillingness to work and wait for wealth which turned them away from commerce. In France, the vanity which is conceded even by Frenchmen to be a national trait led in the same direction. The numbers and brilliancy of the nobility, and the consideration enjoyed by them, set a seal of inferiority upon an occupation which they despised. Rich merchants and manufacturers sighed for the honors of nobility, and upon obtaining them, abandoned their lucrative professions. Therefore, while the industry of the people and the fruitfulness of the soil saved commerce from total decay, it was pursued under a sense of humiliation which caused its best representatives to escape from it as soon as they could. Louis XIV., under the influence of Colbert, put forth an ordinance "authorizing all noblemen to take an interest in merchant ships, goods and merchandise, without being considered as having derogated from nobility, provided they did not sell at retail;" and the reason given for this action was, "that it imports the good of our subjects and our own satisfaction, to efface the relic of a public opinion, universally prevalent, that maritime commerce is incompatible with nobility." But a prejudice involving conscious and open superiority is not readily effaced by ordinances, especially when vanity is a conspicuous trait in national character; and many years later Montesquieu taught that it is contrary to the spirit of monarchy that the nobility should engage in trade.

In Holland there was a nobility; but the State was republican in name, allowed large scope to personal freedom and enterprise, and the centres of power were in the great cities. The foundation of the national greatness was money—or rather wealth. Wealth, as a source of civic distinction, carried with it also power in the State; and with power there went social position and consideration. In England the same result obtained. The nobility were proud; but in a representative government the power of wealth could be neither put down nor overshadowed. It was patent to the eyes of all it was honored by all; and in England, as well as Holland, the occupations which were the source of wealth shared in the honor given to wealth itself. Thus, in all the countries named, social sentiment, the outcome of national characteristics, had a marked influence upon the national attitude toward trade.

In yet another way does the national genius affect the growth of sea power in its broadest sense; and that is in so far as it possesses the capacity for planting healthy colonies. Of colonization, as of all other growths, it is true that it is most healthy when it is most natural. Therefore colonies that spring from the felt wants and natural impulses of a whole people will have the most solid foundations; and their subsequent growth will be surest when they are least trammelled from home, if the people have the genius for independent action. Men of the past three centuries have keenly felt the value to the mother-country of colonies as outlets for the home products and as a nursery for commerce and shipping; but efforts at colonization have not had the same general origin, nor have different systems all had the same success. The efforts of statesmen, however far-seeing and careful, have not been able to supply the lack of strong natural impulse; nor can the most minute regulation from home produce as good results as a happier neglect, when the germ of self-development is found in the national character. There has been no greater display of wisdom in the national administration of successful colonies than in that of unsuccessful. Perhaps there has been even less. If elaborate system and supervision, careful adaptation of means to ends, diligent nursing, could avail for colonial growth, the genius of England has less of this systematizing faculty than the genius of France; but England, not France, has been the great colonizer of the world. Successful colonization, with its consequent effect upon commerce and sea power, depends essentially upon national character; because colonies grow best when they grow of themselves, naturally. The character of the colonist, not the care of the home government, is the principle of the colony's growth.

This truth stands out the clearer because the general attitude of all the home governments toward their colonies was entirely selfish. However founded, as soon as it was recognized to be of consequence, the colony became to the home country a cow to be milked; to be cared for, of course, but chiefly as a piece of property valued for the returns it gave. Legislation was directed toward a monopoly of its external trade; the places in its government afforded posts of value for occupants from the mother-country; and the colony was looked upon, as the sea still so often is, as a fit place for those who were ungovernable or useless at home. The military administration, however, so long as it remains a colony, is the proper and necessary attribute of the home government.

The fact of England's unique and wonderful success as a great colonizing nation is too evident to be dwelt upon; and the reason for it appears to lie chiefly in two traits of the national character. The English colonist naturally and readily settles down in his new country, identifies his interest with it, and though keeping an affectionate remembrance of the home from which he came, has no restless eagerness to return. In the second place, the Englishman at once and instinctively seeks to develop the resources of the new country in the broadest sense. In the former particular he differs from the French, who were ever longingly looking back to the delights of their pleasant land; in the latter, from the Spaniards, whose range of interest and ambition was too narrow for the full evolution of the possibilities of a new country.

The character and the necessities of the Dutch led them naturally to plant colonies; and by the year 1650 they had in the East Indies, in Africa, and in America a large number, only to name which would be tedious. They were then far ahead of England in this matter. But though the origin of these colonies, purely commercial in its character, was natural, there seems to have been lacking to them a principle of growth. "In planting them they never sought an extension of empire, but merely an acquisition of trade and commerce. They attempted conquest only when forced by the pressure of circumstances. Generally they were content to trade under the protection of the sovereign of the country." This placid satisfaction with gain alone, unaccompanied by political ambition, tended, like the despotism of France and Spain, to keep the colonies mere commercial dependencies upon the mother-country, and so killed the natural principle of growth.

Before quitting this head of the inquiry, it is well to ask how far the national character of Americans is fitted to develop a great sea power, should other circumstances become favorable.

It seems scarcely necessary, however, to do more than appeal to a not very distant past to prove that, if legislative hindrances be removed, and more remunerative fields of enterprise filled up, the sea power will not long delay its appearance. The instinct for commerce, bold enterprise in the pursuit of gain, and a keen scent for the trails that lead to it, all exist; and if there be in the future any fields calling for colonization, it cannot be doubted that Americans will carry to them all their inherited aptitude for self-government and independent growth.

6. *Character of the Government.* In discussing the effects upon the development of a nation's sea power exerted by its government and institutions, it will be necessary to avoid a

tendency to over-philosophizing, to confine attention to obvious and immediate causes and their plain results, without prying too far beneath the surface for remote and ultimate influences.

Nevertheless, it must be noted that particular forms of government with their accompanying institutions, and the character of rulers at one time or another, have exercised a very marked influence upon the development of sea power. The various traits of a country and its people which have so far been considered constitute the natural characteristics with which a nation, like a man, begins its career; the conduct of the government in turn corresponds to the exercise of the intelligent will-power, which, according as it is wise, energetic and persevering, or the reverse, causes success or failure in a man's life or a nation's history.

It would seem probable that a government in full accord with the natural bias of its people would most successfully advance its growth in every respect; and, in the matter of sea power, the most brilliant successes have followed where there has been intelligent direction by a government fully imbued with the spirit of the people and conscious of its true general bent. Such a government is most certainly secured when the will of the people, or of their best natural exponents, has some large share in making it; but such free governments have sometimes fallen short, while on the other hand despotic power, wielded with judgment and consistency, has created at times a great sea commerce and a brilliant navy with greater directness than can be reached by the slower processes of a free people. The difficulty in the latter case is to insure perseverance after the death of a particular despot.

England having undoubtedly reached the greatest height of sea power of any modern nation, the action of her government first claims attention. In general direction this action has been consistent, though often far from praiseworthy. It has aimed steadily at the control of the sea. One of its most arrogant expressions dates back as far as the reign of James I., when she had scarce any possessions outside her own islands; before Virginia or Massachusetts was settled. Here is Richelieu's account of it:

> The Duke of Sully, minister of Henry IV. [one of the most chivalrous princes that ever lived], having embarked at Calais in a French ship wearing the French flag at the main, was no sooner in the Channel than, meeting an English despatch-boat which was there to receive him, the commander of the latter ordered the French ship to lower her flag. The Duke, considering that his quality freed him from such an affront, boldly refused; but this refusal was followed by three cannon-shot, which, piercing his ship, pierced the heart likewise of all good Frenchmen. Might forced him to yield what right forbade, and for all the complaints he made he could get no better reply from the English captain than this: 'That just as his duty obliged him to honor the ambassador's rank, it also obliged him to exact the honor due to the flag of his master as sovereign of the sea.' If the words of King James himself were more polite, they nevertheless had no other effect than to compel the Duke to take counsel of his prudence, feigning to be satisfied, while his wound was all the time smarting and incurable. Henry the Great had to practise moderation on this occasion; but with the resolve another time to sustain the rights of his crown by the force that, with the aid of time, he should be able to put upon the sea.

This act of unpardonable insolence, according to modern ideas, was not so much out of accord with the spirit of nations in that day. It is chiefly noteworthy as the most striking, as well as one of the earliest indications of the purpose of England to assert herself at all risks upon the sea; and the insult was offered under one of her most timid kings to an ambassador immediately representing the bravest and ablest of French sovereigns. This empty honor of the flag, a claim insignificant except as the outward manifestation of the purpose of a government, was as rigidly exacted under Cromwell as under the kings. It was one of the conditions of peace yielded by the Dutch after their disastrous war of 1654. Cromwell, a despot in everything but name, was keenly alive to all that concerned England's honor and strength, and did not stop at barren salutes to promote them. Hardly yet possessed of power, the English navy sprang rapidly into a new life and vigor under his stern rule. England's rights, or reparation for her wrongs, were demanded by her fleets throughout the world,—in the Baltic, in the Mediterranean, against the Barbary States, in the West Indies; and under him the conquest of Jamaica began that extension of her empire, by force of arms, which has gone on to our own days. Nor were equally strong peaceful measures for the growth of English trade and shipping forgotten. Cromwell's celebrated Navigation Act declared that all imports into England or her colonies must be conveyed exclusively in vessels belonging to England herself, or to the country in which the products carried were grown or manufactured. This decree, aimed specially at the Dutch, the common carriers of Europe, was resented throughout the commercial world; but the benefit to England, in those days of national strife and animosity, was so apparent that it lasted long under the monarchy. A century and a quarter later we find Nelson, before his famous career had begun, showing his zeal for the welfare of England's shipping by enforcing this same act in the West Indies against American merchant-ships. When Cromwell was dead, and Charles II. sat on the throne of his father, this king, false to the English people, was yet true to England's greatness and to the traditional policy of her government on the sea. In his treacherous intrigues with Louis XIV., by which he aimed to make himself independent of Parliament and people, he wrote to Louis: "There are two impediments to a perfect union. The first is the great care France is now taking to create a commerce and to be an imposing maritime power. This is so great a cause of suspicion with us, who can possess importance only by our commerce and our naval force, that every step which France takes in this direction will perpetuate the jealousy between the two nations." In the midst of the negotiations which preceded the detestable attack of the two kings upon the Dutch republic, a warm dispute arose as to who should command the united fleets of France and England. Charles was inflexible on this point. "It is the custom of the English," said he, "to command at sea;" and he told the French ambassador plainly that, were he to yield, his subjects would not obey him. In the projected partition of the United Provinces he reserved for England the maritime plunder in positions that controlled the mouths of the rivers Scheldt and Meuse. The navy under Charles preserved for some time the spirit and discipline impressed on it by Cromwell's iron rule; though later it shared in the general decay of *morale* which marked this evil reign. Monk, having by a great strategic blunder sent off a fourth of his fleet, found himself in 1666 in presence of a greatly superior Dutch force. Disregarding the odds, he attacked without hesitation, and for three days maintained the fight with honor, though with loss. Such conduct is not war; but in the single eye that looked to England's naval prestige and dictated his action, common as it was to England's people as well as to her government, has lain the secret of final success following many blunders through the centuries. Charles's successor, James II., was himself a seaman, and had commanded in two great sea-fights. When William III. came to the throne, the governments of England and Holland were under one hand, and continued united in one purpose against Louis XIV. until the Peace of Utrecht in 1713; that is, for a quarter of a century. The English government more and more steadily, and with conscious purpose, pushed on the extension of her sea dominion and fostered the growth of her sea power. While as an open enemy she struck at France upon the sea, so as an artful friend, many at least believed, she sapped the power of Holland afloat. The treaty between the two countries provided that of the sea forces Holland should furnish three eighths, England five eighths, or nearly double. Such a provision, coupled with a further one which made Holland keep up an army of 102,000 against England's 40,000, virtually threw the land war on one and the sea war on the other. The tendency, whether designed or not, is evident; and at the peace, while Holland received compensation by land, England obtained, besides commercial privileges in France, Spain, and the Spanish West Indies, the important maritime concessions of Gibraltar and Port Mahon in the Mediterranean; of Newfoundland, Nova Scotia, and Hudson's Bay in North America. The naval power of France and Spain had disappeared; that of Holland thenceforth steadily declined. Posted thus in America, the West Indies, and the Mediterranean, the English government thenceforth moved firmly forward on the path which made of the English kingdom the British Empire. For the twenty-five years following the Peace of Utrecht, peace was the chief aim of the ministers who directed the policy of the two great seaboard nations, France and England; but amid all the fluctuations of continental politics in a most unsettled period, abounding in petty wars and shifty treaties, the eye of England was steadily fixed on the maintenance of her sea power. In the Baltic, her fleets checked the attempts of Peter the Great upon Sweden, and so maintained a balance of power in that sea, from which she drew not only a great trade but the chief part of her naval stores, and which the Czar aimed to make a Russian lake. Denmark endeavored to establish an East India company aided by foreign capital; England and Holland not only forbade their subjects to join it, but threatened Denmark, and thus stopped an enterprise they thought adverse to their sea interests. In the Netherlands, which by the Utrecht Treaty had passed to Austria, a similar East India company, having Ostend for its port, was formed, with the emperor's sanction. This step, meant to restore to the Low Countries the trade lost to them through their natural outlet of the Scheldt, was opposed by the sea powers England and Holland; and their greediness for the monopoly of trade, helped in this instance by France, stifled this company also after a few years of struggling life. In the Mediterranean, the Utrecht settlement was disturbed by the emperor of Austria, England's natural ally in the then existing state of European politics. Backed by England, he, having already Naples, claimed also Sicily in exchange for Sardinia. Spain resisted; and her navy, just beginning to revive under a vigorous minister, Alberoni, was crushed and annihilated by the English fleet off Cape Passaro in 1718; while the following year a French army, at the bidding of England, crossed the Pyrenees and completed the work by destroying the Spanish dock-yards. Thus England, in addition to Gibraltar and Mahon in her own hands, saw Naples and Sicily in those of a friend, while an enemy was struck down. In Spanish America, the limited privileges to English trade, wrung from the necessities of Spain, were abused by an extensive and scarcely disguised smuggling system; and when the exasperated Spanish government gave way to excesses in the mode of suppression, both the minister who counselled peace and the opposition which urged war defended their opinions by alleging the effects of either upon England's sea power and honor. While England's policy thus steadily aimed at widening and strengthening the bases of her sway upon the ocean, the other governments of Europe seemed blind to the dangers to be feared from her sea growth. The miseries resulting from the overweening power of Spain in days long gone by seemed to be forgotten; forgotten also the more recent lesson of the bloody and costly wars provoked by the ambition and exaggerated power of Louis XIV. Under the eyes of the statesmen of Europe there was steadily and visibly being built up a third overwhelming power, destined to be used as selfishly, as aggressively, though not as cruelly, and much more successfully than any that had preceded it. This was the power of the sea, whose workings, because more silent than the clash of arms, are less often noted, though lying clearly enough on the surface. It can scarcely be denied that England's uncontrolled dominion of the seas, during almost the whole period chosen for our subject, was by long odds the chief among the military factors that determined the final issue.[2] So far, however, was this influence from being foreseen after Utrecht, that France for twelve years, moved by personal exigencies of her rulers, sided with England against Spain; and when Fleuri came into power in 1726, though this policy was reversed, the navy of France received no attention, and the only blow at England was the establishment of a Bourbon prince, a natural enemy to her, upon the throne of the two Sicilies in 1736. When war broke out with Spain in 1739, the navy of England was in numbers more than equal to the combined navies of Spain and France; and during the quarter of a century of nearly uninterrupted war that followed, this numerical disproportion increased. In these wars England, at first instinctively, afterward with conscious purpose under a government that recognized her opportunity and the possibilities of her great sea power, rapidly built up that mighty colonial empire whose foundations were already securely laid in the characteristics of her colonists and the strength of her fleets. In strictly European affairs her wealth, the outcome of

2. An interesting proof of the weight attributed to the naval power of Great Britain by a great military authority will be found in the opening chapter of Jomini's "History of the Wars of the French Revolution." He lays down, as a fundamental principle of European policy, than an unlimited expansion of naval force should not be permitted to any nation which cannot be approached by land,—a description which can apply only to Great Britain.

her sea power, made her play a conspicuous part during the same period. The system of subsidies, which began half a century before in the wars of Marlborough and received its most extensive development half a century later in the Napoleonic wars, maintained the efforts of her allies, which would have been crippled, if not paralyzed, without them. Who can deny that the government which with one hand strengthened its fainting allies on the continent with the life-blood of money, and with the other drove its own enemies off the sea and out of their chief possessions, Canada, Martinique, Guadeloupe, Havana, Manila, gave to its country the foremost rôle in European politics; and who can fail to see that the power which dwelt in that government, with a land narrow in extent and poor in resources, sprang directly from the sea? The policy in which the English government carried on the war is shown by a speech of Pitt, the master-spirit during its course, though he lost office before bringing it to an end. Condemning the Peace of 1763, made by his political opponent, he said: "France is chiefly, if not exclusively, formidable to us as a maritime and commercial power. What we gain in this respect is valuable to us, above all, through the injury to her which results from it. You have left to France the possibility of reviving her navy." Yet England's gains were enormous; her rule in India was assured, and all North America east of the Mississippi in her hands. By this time the onward path of her government was clearly marked out, had assumed the force of a tradition, and was consistently followed. The war of the American Revolution was, it is true, a great mistake, looked at from the point of view of sea power; but the government was led into it insensibly by a series of natural blunders. Putting aside political and constitutional considerations, and looking at the question as purely military or naval, the case was this: The American colonies were large and growing communities at a great distance from England. So long as they remained attached to the mother-country, as they then were enthusiastically, they formed a solid base for her sea power in that part of the world; but their extent and population were too great, when coupled with the distance from England, to afford any hope of holding them by force, *if* any powerful nations were willing to help them. This "if," however, involved a notorious probability; the humiliation of France and Spain was so bitter and so recent that they were sure to seek revenge, and it was well known that France in particular had been carefully and rapidly building up her navy. Had the colonies been thirteen islands, the sea power of England would quickly have settled the question; but instead of such a physical barrier they were separated only by local jealousies which a common danger sufficiently overcame. To enter deliberately on such a contest, to try to hold by force so extensive a territory, with a large hostile population, so far from home, was to renew the Seven Years' War with France and Spain, and with the Americans, against, instead of for, England. The Seven Years' War had been so heavy a burden that a wise government would have known that the added weight could not be borne, and have seen it was necessary to conciliate the colonists. The government of the day was not wise, and a large element of England's sea power was sacrificed; but by mistake, not wilfully; through arrogance, not through weakness.

This steady keeping to a general line of policy was doubtless made specially easy for successive English governments by the clear indications of the country's conditions. Singleness of purpose was to some extent imposed. The firm maintenance of her sea power, the haughty determination to make it felt, the wise state of preparation in which its military element was kept, were yet more due to that feature of her political institutions which practically gave the government during the period in question, into the hands of a class,—a landed aristocracy. Such a class, whatever its defects otherwise, readily takes up and carries on a sound political tradition, is naturally proud of its country's glory, and comparatively insensible to the sufferings of the community by which that glory is maintained. It readily lays on the pecuniary burden necessary for preparation and for endurance of war. Being as a body rich, it feels those burdens less. Not being commercial, the sources of its own wealth are not so immediately endangered, and it does not share that political timidity which characterizes those whose property is exposed and business threatened,—the proverbial timidity of capital. Yet in England this class was not insensible to anything that touched her trade for good or ill. Both houses of Parliament vied in careful watchfulness over its extension and protection, and to the frequency of their inquiries a naval historian attributes the increased efficiency of the executive power in its management of the navy. Such a class also naturally imbibes and keeps up a spirit of military honor, which is of the first importance in ages when military institutions have not yet provided the sufficient substitute in what is called *esprit-de-corps*. But although full of class feeling and class prejudice, which made themselves felt in the navy as well as elsewhere, their practical sense left open the way of promotion to its highest honors to the more humbly born; and every age saw admirals who had sprung from the lowest of the people. In this the temper of the English upper class differed markedly from that of the French. As late as 1789, at the outbreak of the Revolution, the French Navy List still bore the name of an official whose duty was to verify the proofs of noble birth on the part of those intending to enter the naval school.

Since 1815, and especially in our own day, the government of England has passed very much more into the hands of the people at large. Whether her sea power will suffer therefrom remains to be seen. Its broad basis still remains in a great trade, large mechanical industries, and an extensive colonial system. Whether a democratic government will have the foresight, the keen sensitiveness to national position and credit, the willingness to insure its prosperity by adequate outpouring of money in times of peace, all which are necessary for military preparation, is yet an open question. Popular governments are not generally favorable to military expenditure, however necessary, and there are signs that England tends to drop behind.

It has already been seen that the Dutch Republic, even more that the English nation, drew its prosperity and its very life from the sea. The character and policy of its government were far less favorable to a consistent support of sea power. Composed of seven provinces, with the political name of the United Provinces, the actual distribution of power may be roughly described to Americans as an exaggerated example of States Rights. Each of the maritime provinces had its own fleet and its own admiralty, with consequent jealousies. This disorganizing tendency was partly counteracted by the great preponderance of the Province of Holland, which alone contributed five sixths of the fleet and fifty-eight per cent of the taxes, and consequently had a proportionate share in directing the national policy. Although intensely patriotic, and capable of making the last sacrifices for freedom, the commercial spirit of the people penetrated the government, which indeed might be called a commercial aristocracy, and made it averse to war, and to the expenditures which are necessary in preparing for war. As has before been said, it was not until danger stared them in the face that the burgomasters were willing to pay for their defences. While the republican government lasted, however, this economy was practised least of all upon the fleet; and until the death of John De Witt, in 1672, and the peace with England in 1674, the Dutch navy was in point of numbers and equipment able to make a fair show against the combined navies of England and France. Its efficiency at this time undoubtedly saved the country from the destruction planned by the two kings. With De Witt's death the republic passed away, and was followed by the practically monarchical government of William of Orange. The life-long policy of this prince, then only eighteen, was resistance to Louis XIV and to the extension of French power. This resistance took shape upon the land rather than the sea,—a tendency promoted by England's withdrawal from the war. As early as 1676, Admiral De Ruyter found the force given him unequal to cope with the French alone. With the eyes of the government fixed on the land frontier, the navy rapidly declined. In 1688, when William of Orange needed a fleet to convoy him to England, the burgomasters of Amsterdam objected that the navy was incalculably decreased in strength, as well as deprived of its ablest commanders. When king of England, William still kept his position as stadtholder, and with it his general European policy. He found in England the sea power he needed, and used the resources of Holland for the land war. This Dutch prince consented that in the allied fleets, in councils of war, the Dutch admirals should sit below the junior English captain; and Dutch interests at sea were sacrificed as readily as Dutch pride to the demands of England. When William died, his policy was still followed by the government which succeeded him. Its aims were wholly centred upon the land, and at the Peace of Utrecht, which closed a series of wars extending over forty years, Holland, having established no sea claim, gained nothing in the way of sea resources, of colonial extension, or of commerce.

Of the last of these wars an English historian says: "The economy of the Dutch greatly hurt their reputation and their trade. Their men-of-war in the Mediterranean were always victualled short, and their convoys were so weak and ill-provided that for one ship that we lost, they lost five, which begat a general notion that we were the safer carriers, which certainly had a good effect. Hence it was that our trade rather increased than diminished in this war."

From that time Holland ceased to have a great sea power, and rapidly lost the leading position among the nations which that power had built up. It is only just to say that no policy could have saved from decline this small, though determined, nation, in face of the persistent enmity of Louis XIV. The friendship of France, insuring peace on her landward frontier, would have enabled her, at least for a longer time, to dispute with England the dominion of the seas; and as allies the navies of the two continental States might have checked the growth of the enormous sea power which has just been considered. Sea peace between England and Holland was only possible by the virtual subjection of one or the other, for both aimed at the same object. Between France and Holland it was otherwise; and the fall of Holland proceeded, not necessarily from her inferior size and numbers, but from faulty policy on the part of the two governments. It does not concern us to decide which was the more to blame.

France, admirably situated for the possession of sea power, received a definite policy for the guidance of her government from two great rulers, Henry IV. and Richelieu. With certain well-defined projects of extension eastward upon the land were combined a steady resistance to the House of Austria, which then ruled in both Austria and Spain, and an equal purpose of resistance to England upon the sea. To further this latter end, as well as for other reasons, Holland was to be courted as an ally. Commerce and fisheries as the basis of sea power were to be encouraged, and a military navy was to be built up. Richelieu left what he called his political will, in which he pointed out the opportunities of France for achieving sea power, based upon her position and resources; and French writers consider him the virtual founder of the navy, not merely because he equipped ships, but from the breadth of his views and his measures to insure sound institutions and steady growth. After his death, Mazarin inherited his views and general policy, but not his lofty and martial spirit, and during his rule the newly formed navy disappeared. When Louis XIV. took the government into his own hands, in 1661, there were but thirty ships of war, of which only three had as many as sixty guns. Then began a most astonishing manifestation of the work which can be done by absolute government ably and systematically wielded. That part of the administration which dealt with trade, manufactures, shipping, and colonies, was given to a man of great practical genius, Colbert, who had served with Richelieu and had drunk in fully his ideas and policy. He pursued his aims in a spirit thoroughly French. Everything was to be organized, the spring of everything was in the minister's cabinet. "To organize producers and merchants as a powerful army, subjected to an active and intelligent guidance, so as to secure an industrial victory for France by order and unity of efforts, and to obtain the best products by imposing on all workmen the processes recognized as best by competent men. . . . To organize seamen and distant commerce in large bodies like the manufactures and internal commerce, and to give as a support to the commercial power of France a navy established on a firm basis and of dimensions hitherto unknown,"—such, we are told, were the aims of Colbert as regards two of the three links in the chain of sea power. For the third, the colonies at the far end of the line, the same governmental direction and organization were evidently purposed; for the government began by buying back Canada, Newfoundland, Nova Scotia, and the French West India Islands from the parties who then owned them. Here, then, is seen pure, absolute, uncontrolled power gathering up into its hands all the reins for the guidance of a nation's course, and proposing so to direct it as to make, among other things, a great sea power.

To enter into the details of Colbert's action is beyond our purpose. It is enough to note the chief part played by the government in building up the sea power of the State, and that this very great man looked not to any one of the bases on which it rests to the exclusion of the others, but embraced them all in his wise and provident administration. Agriculture, which increases the products of the earth, and manufactures, which multiply the products of man's industry; internal trade routes and regulations, by which the exchange of products from the interior to the exterior is made easier; shipping and customs regulations tending to throw the carrying-trade into French hands, and so to encourage the building of French shipping, by which the home and colonial products should be carried back and forth; colonial administration and development, by which a far-off market might be continually growing up to be monopolized by the home trade; treaties with foreign States favoring French trade, and imposts on foreign ships and products tending to break down that of rival nations,—all these means, embracing countless details, were employed to build up for France (1) Production; (2) Shipping; (3) Colonies and Markets,—in a word, sea power. The study of such a work is simpler and easier when thus done by one man, sketched out by a kind of logical process, than when slowly wrought by conflicting interests in a more complex government. In the few years of Colbert's administration is seen the whole theory of sea power put into practice in the systematic, centralizing French way; while the illustration of the same theory in English and Dutch history is spread over generations. Such growth, however, was forced, and depended upon the endurance of the absolute power which watched over it; and as Colbert was not king, his control lasted only till he lost the king's favor. It is, however, most interesting to note the results of his labors in the proper field for governmental action—in the navy. It has been said that in 1661, when he took office, there were but thirty armed ships, of which three only had over sixty guns. In 1666 there were seventy, of which fifty were ships of the line and twenty were fire-ships; in 1671, from seventy the number had increased to one hundred and ninety-six. In 1683 there were one hundred and seven ships of from twenty-four to one hundred and twenty guns, twelve of which carried over seventy-six guns, besides many smaller vessels. The order and system introduced into the dock-yards made them vastly more efficient than the English. An English captain, a prisoner in France while the effect of Colbert's work still lasted in the hands of his sons, writes:

> When I was first brought prisoner thither, I lay four months in a hospital at Brest for care of my wounds. While there I was astonished at the expedition used in manning and fitting out their ships, which till then I thought could be done nowhere sooner than in England, where we have ten times the shipping, and consequently ten times the seamen, they have in France; but there I saw twenty sail of ships, of about sixty guns each, got ready in twenty days' time; they were brought in and the men were discharged; and upon an order from Paris they were careened, keeled up, rigged, victualled, manned, and out again in the said time with the greatest ease imaginable. I likewise saw a ship of one hundred guns that had all her guns taken out in four or five hours' time; which I never saw done in England in twenty-four hours, and this with the greatest ease and less hazard than at home. This I saw under my hospital window.

A French naval historian cites certain performances which are simply incredible, such as that the keel of a galley was laid at four o'clock, and that at nine she left port, fully armed. These traditions may be accepted as pointing, with the more serious statements of the English officer, to a remarkable degree of system and order, and abundant facilities for work.

Yet all this wonderful growth, forced by the action of the government, withered away like Jonah's gourd when the government's favor was withdrawn. Time was not allowed for its roots to strike down deep into the life of the nation. Colbert's work was in the direct line of Richelieu's policy, and for a time it seemed there would continue the course of action which would make France great upon the sea as well as predominant upon the land. For reasons which it is not yet necessary to give, Louis came to have feelings of bitter enmity against Holland; and as these feelings were shared by Charles II., the two kings determined on the destruction of the United Provinces. This war, which broke out in 1672, though more contrary to natural feeling on the part of England, was less of a political mistake for her than for France, and especially as regards sea power. France was helping to destroy a probable, and certainly an indispensable, ally; England was assisting in the ruin of her greatest rival on the sea, at this time, indeed, still her commercial superior. France, staggering under debt and utter confusion in her finances when Louis mounted the throne, was just seeing her way clear in 1672, under Colbert's reforms and their happy results. The war, lasting six years, undid the greater part of his work. The agricultural classes, manufactures, commerce, and the colonies, all were smitten by it; the establishments of Colbert languished, and the order he had established in the finances was overthrown. Thus the action of Louis—and he alone was the directing government of France—struck at the roots of her sea power, and alienated her best sea ally. The territory and the military power of France were increased, but the springs of commerce and of a peaceful shipping had been exhausted in the process; and although the military navy was for some years kept up with splendor and efficiency, it soon began to dwindle, and by the end of the reign had practically disappeared. The same false policy, as regards the sea, marked the rest of this reign of fifty-four years. Louis steadily turned his back upon the sea interests of France, except the fighting-ships, and either could not or would not see that the latter were of little use and uncertain life, if the peaceful shipping and the industries, by which they were supported, perished. His policy, aiming at supreme power in Europe by military strength and territorial extension, forced England and Holland into an alliance, which, as has before been said, directly drove France off the sea, and indirectly swamped Holland's power thereon. Colbert's navy perished, and for the last ten years of Louis' life no great French fleet put to sea, though there was constant war. The simplicity of form in an absolute monarchy thus brought out strongly how great the influence of government can be upon both the growth and the decay of sea power.

The latter part of Louis' life thus witnessed that power failing by the weakening of its foundations, of commerce, and of the wealth that commerce brings. The government that followed, likewise absolute, of set purpose and at the demand of England, gave up all pretence of maintaining an effective navy. The reason for this was that the new king was a minor; and the regent, being bitterly at enmity with the king of Spain, to injure him and preserve his own power, entered into alliance with England. He aided her to establish Austria, the hereditary enemy of France, in Naples and Sicily to the detriment of Spain, and in union with her destroyed the Spanish navy and dock-yards. Here again is found a personal ruler disregarding the sea interests of France, ruining a natural ally, and directly aiding, as Louis XIV. indirectly and unintentionally aided, the growth of a mistress of the seas. This transient phase of policy passed away with the death of the regent in 1726; but from that time until 1760 the government of France continued to disregard her maritime interests. It is said, indeed, that owing to some wise modifications of her fiscal regulations, mainly in the direction of free trade (and due to Law, a minister of Scotch birth), commerce with the East and West Indies wonderfully increased, and that the islands of Guadeloupe and Martinique became very rich and thriving; but both commerce and colonies lay at the mercy of England when war came, for the navy fell into decay. In 1756, when things were no longer at their worst, France had but forty-five ships-of-the-line, England nearly one hundred and thirty; and when the forty-five were to be armed and equipped, there was found to be neither material nor rigging nor supplies; not even enough artillery. Nor was this all.

> "Lack of system in the government," says a French writer, "brought about indifference, and opened the door to disorder and lack of discipline. Never had unjust promotions been so frequent; so also never had more universal discontent been seen. Money and intrigue took the place of all else, and brought in their train commands and power. Nobles and upstarts, with influence at the capital and self-sufficiency in the seaports, thought themselves dispensed with merit. Waste of the revenues of the State and of the dock-yards knew no bounds. Honor and modesty were turned into ridicule. As if the evils were not thus great enough, the ministry took pains to efface the heroic traditions of the past which had escaped the general wreck. To the energetic fights of the great reign succeeded, by order of the court, 'affairs of circumspection.' To preserve to the wasted material a few armed ships, increased opportunity was given to the enemy. From this unhappy principle we were bound to a defensive as advantageous to the enemy as it was foreign to the genius of our people. This circumspection before the enemy, laid down for us by orders, betrayed in the long run the national temper; and the abuse of the system led to acts of indiscipline and defection under fire, of which a single instance would vainly be sought in the previous century."

A false policy of continental extension swallowed up the resources of the country, and was doubly injurious because, by leaving defenceless its colonies and commerce, it exposed the greatest source of wealth to be cut off, as in fact happened. The small squadrons that got to sea were destroyed by vastly superior force; the merchant shipping was swept away, and the colonies, Canada, Martinique, Guadeloupe, India, fell into England's hands. If it did not take too much space, interesting extracts might be made, showing the woful misery of France, the country that had abandoned the sea, and the growing wealth of England amid all her sacrifices and exertions. A contemporary writer has thus expressed his view of the policy of France at this period:

> France, by engaging so heartily as she has done in the German war, has drawn away so much of her attention and her revenue from her navy that it enabled us to give such a blow to her maritime strength as possibly she may never be able to recover. Her engagement in the German war has likewise drawn her from the defence of her colonies, by which means we have conquered some of the most considerable she possessed. It has withdrawn her from the protection of her trade, by which it is entirely destroyed, while that of England has never, in the profoundest peace, been in so flourishing a condition. So that, by embarking in this German war, France has suffered herself to be undone, so far as regards her particular and immediate quarrel with England.

In the Seven Years' War France lost thirty-seven ships-of-the-line and fifty-six frigates,—a force three times as numerous as the whole navy of the United States at any time in the days of sailing-ships. "For the first time since the Middle Ages," says a French historian, speaking of the same war, "England had conquered France single-handed, almost without allies, France having powerful auxiliaries. She had conquered solely by the superiority of her government." Yes; but it was by the superiority of her government using the tremendous weapon of her sea power,—the reward of a consistent policy perseveringly directed to one aim.

The profound humiliation of France, which reached its depths between 1760 and 1763, at which latter date she made peace, has an instructive lesson for the United States in this our period of commercial and naval decadence. We have been spared her humiliation; let us hope to profit by her subsequent example. Between the same years (1760 and 1763) the French people rose, as afterward in 1793, and declared they would have a navy. "Popular feeling, skilfully directed by the government, took up the cry from one end of France to the other, 'The navy must be restored.' Gifts of ships were made by cities, by corporations, and by private subscriptions. A prodigious activity sprang up in the lately silent ports; everywhere ships were building or repairing." This activity was sustained; the arsenals were replenished, the material of every kind was put on a satisfactory footing, the artillery reorganized, and ten thousand trained gunners drilled and maintained.

The tone and action of the naval officers of the day instantly felt the popular impulse, for which indeed some loftier spirits among them had been not only waiting but working. At no time was greater mental and professional activity found among French naval officers than just then, when their ships had been suffered to rot away by governmental inaction. Thus a prominent French officer of our own day writes:

> The sad condition of the navy in the reign of Louis XV., by closing to officers the brilliant career of bold enterprises and successful battles, forced them to fall back upon themselves. They drew from study the knowledge they were to put to the proof some years later, thus putting into practice that fine saying of Montesquieu, 'Adversity is our mother, Prosperity our step-mother.' . . . By the year 1769 was seen in all its splendor that brilliant galaxy of officers whose activity stretched to the ends of the earth, and who embraced in their works and in their investigations all the branches of human knowledge. The Académie de Marine, founded in 1752, was reorganized.[3]

3. A. Gougeard: *La Marine de Guerre: Richelieu et Colbert* (1877).

The Académie's first director, a post-captain named Bigot de Morogues, wrote an elaborate treatise on naval tactics, the first original work on the subject since Paul Hoste's, which it was designed to supersede. Morogues must have been studying and formulating his problems in tactics in days when France had no fleet, and was unable so much as to raise her head at sea under the blows of her enemy. At the same time England had no similar book; and an English lieutenant, in 1762, was just translating a part of Hoste's great work, omitting by far the larger part. It was not until nearly twenty years later that Clerk, a Scotch private gentleman, published an ingenious study of naval tactics, in which he pointed out to English admirals the system by which the French had thwarted their thoughtless and ill-combined attacks.[4] "The researches of the Académie de Marine, and the energetic impulse which it gave to the labors of officers, were not, as we hope to show later, without influence upon the relatively prosperous condition in which the navy was at the beginning of the American war."

It has already been pointed out that the American War of Independence involved a departure from England's traditional and true policy, by committing her to a distant land war, while powerful enemies were waiting for an opportunity to attack her at sea. Like France in the then recent German wars, like Napoleon later in the Spanish war, England, through undue self-confidence, was about to turn a friend into an enemy, and so expose the real basis of her power to a rude proof. The French government, on the other hand, avoided the snare into which it had so often fallen. Turning

4. Whatever may be thought of Clerk's claim to originality in constructing a system of naval tactics, and it has been seriously impugned, there can be no doubt that his criticisms on the past were sound. So far as the author knows, he in this respect deserves credit for an originality remarkable in one who had the training neither of a seaman nor of a military man.

her back on the European continent, having the probability of neutrality there, and the certainty of alliance with Spain by her side, France advanced to the contest with a fine navy and a brilliant, though perhaps relatively inexperienced, body of officers. On the other side of the Atlantic she had the support of a friendly people, and of her own or allied ports, both in the West Indies and on the continent. The wisdom of this policy, the happy influence of this action of the government upon her sea power, is evident; but the details of the war do not belong to this part of the subject. To Americans, the chief interest of that war is found upon the land; but to naval officers upon the sea, for it was essentially a sea war. The intelligent and systematic efforts of twenty years bore their due fruit; for though the warfare afloat ended with a great disaster, the combined efforts of the French and Spanish fleets undoubtedly bore down England's strength and robbed her of her colonies. In the various naval undertakings and battles the honor of France was upon the whole maintained; though it is difficult, upon consideration of the general subject, to avoid the conclusion that the inexperience of French seamen as compared with English, the narrow spirit of jealousy shown by the noble corps of officers toward those of different antecedents, and above all, the miserable traditions of three quarters of a century already alluded to, the miserable policy of a government which taught them first to save their ships, to economize the material, prevented French admirals from reaping, not the mere glory, but the positive advantages that more than once were within their grasp. When Monk said the nation that would rule upon the sea must always attack, he set the key-note to England's naval policy; and had the instructions of the French government consistently breathed the same spirit, the war of 1778 might have ended sooner and better than it did. It seems ungracious to criticise the conduct of a service to which, under God, our nation owes that its birth was not a miscarriage; but writers of its own country abundantly reflect the spirit of the remark. A French officer who served afloat during this war, in a work of calm and judicial tone, says:

> What must the young officers have thought who were at Sandy Hook with D'Estaing, at St. Christopher with De Grasse, even those who arrived at Rhode Island with De Ternay, when they saw that these officers were not tried at their return?[5]

Again, another French officer, of much later date, justifies the opinion expressed, when speaking of the war of the American Revolution in the following terms:

> It was necessary to get rid of the unhappy prejudices of the days of the regency and of Louis XV.; but the mishaps of which they were full were too recent to be forgotten by our ministers. Thanks to a wretched hesitation, fleets, which had rightly alarmed England, became reduced to ordinary proportions. Intrenching themselves in a false economy, the ministry claimed that, by reason of the excessive expenses necessary to maintain the fleet, the admirals must be ordered to maintain the '*greatest circumspection,*' as though in war half measures have not always led to disasters. So, too, the orders given to our squadron chiefs were to keep the sea as long as possible, without engaging in actions which might cause the loss of vessels difficult to replace; so that more than once complete victories, which would have crowned the skill of our admirals and the courage of our captains, were changed into successes of little importance. A system which laid down as a principle that an admiral should not use the force in his hands, which sent him against the enemy with the fore-ordained purpose of receiving rather than making the attack, a system which sapped moral power to save material resources, must have unhappy results. . . . It is certain that this deplorable system was one of the causes of the lack of discipline and startling

5. Laserre, *Essais historique et critique de la marine française.*

defections which marked the periods of Louis XVI., of the [first] Republic, and of the [first] Empire.[6]

Within ten years of the peace of 1783 came the French Revolution; but that great upheaval which shook the foundations of States, loosed the ties of social order, and drove out of the navy nearly all the trained officers of the monarchy who were attached to the old state of things, did not free the French navy from a false system. It was easier to overturn the form of government than to uproot a deep-seated tradition. Hear again a third French officer, of the highest rank and literary accomplishments, speaking of the inaction of Villeneuve, the admiral who commanded the French rear at the battle of the Nile, and who did not leave his anchors while the head of the column was being destroyed:

> A day was to come [Trafalgar] in which Villeneuve in his turn, like De Grasse before him, and like Duchayla, would complain of being abandoned by part of his fleet. We have come to suspect some secret reason for this fatal coincidence. It is not natural that among so many honorable men there should so often be found admirals and captains incurring such a reproach. If the name of some of them is to this very day sadly associated with the memory of our disasters, we may be sure the fault is not wholly their own. We must rather blame the nature of the operations in which they were engaged, and that system of defensive war prescribed by the French government, which Pitt, in the English Parliament, proclaimed to be the forerunner of certain ruin. That system, when we wished to renounce it, had already penetrated our habits; it had, so to say, weakened our arms and paralyzed our self-reliance. Too often did our squadrons leave port with a special mission to fulfil, and with the intention of avoiding the enemy; to fall in with him was at once a piece of back luck. It was thus that our ships went into action; they submitted to it instead of forcing it. . . . Fortune would have hesitated longer between

6. Lapeyrouse-Bonfils, *Histoire de la marine française.*

> the two fleets, and not have borne in the end so heavily against ours, if Brueys, meeting Nelson half way, could have gone out to fight him. This fettered and timid war, which Villaret and Martin had carried on, had lasted long, thanks to the circumspection of some English admirals and the traditions of the old tactics. It was with these traditions that the battle of the Nile had broken; the hour for decisive action had come.[7]

Some years later came Trafalgar, and again the government of France took up a new policy with the navy. The author last quoted speaks again:

> The emperor, whose eagle glance traced plans of campaign for his fleets as for his armies, was wearied by these unexpected reverses. He turned his eyes from the one field of battle in which fortune was faithless to him, and decided to pursue England elsewhere than upon the seas; he undertook to rebuild his navy, but without giving it any part in the struggle which became more furious than ever. . . . Nevertheless, far from slackening, the activity of our dock-yards redoubled. Every year ships-of-the-line were either laid down or added to the fleet. Venice and Genoa, under his control, saw their old splendors rise again, and from the shores of the Elbe to the head of the Adriatic all the ports of the continent emulously seconded the creative thought of the emperor. Numerous squadrons were assembled in the Scheldt, in Brest Roads, and in Toulon. . . . But to the end the emperor refused to give this navy, full of ardor and self-reliance, an opportunity to measure its strength with the enemy. . . . Cast down by constant reverses, he had kept up our armed ships only to oblige our enemies to blockades whose enormous cost must end by exhausting their finances.

When the empire fell, France had one hundred and three ships-of-the-line and fifty-five frigates.

To turn now from the particular lessons drawn from the

7. Jurien de la Gravière, *Guerres Maritimes*.

history of the past to the general question of the influence of government upon the sea career of its people, it is seen that that influence can work in two distinct but closely related ways.

First, in peace: The government by its policy can favor the natural growth of a people's industries and its tendencies to seek adventure and gain by way of the sea; or it can try to develop such industries and such sea-going bent, when they do not naturally exist; or, on the other hand, the government may by mistaken action check and fetter the progress which the people left to themselves would make. In any one of these ways the influence of the government will be felt, making or marring the sea power of the country in the matter of peaceful commerce; upon which alone, it cannot be too often insisted, a thoroughly strong navy can be based.

Secondly, for war: The influence of the government will be felt in its most legitimate manner in maintaining an armed navy, of a size commensurate with the growth of its shipping and the importance of the interests connected with it. More important even than the size of the navy is the question of its institutions, favoring a healthful spirit and activity, and providing for rapid development in time of war by an adequate reserve of men and of ships and by measures for drawing out that general reserve power which has before been pointed to, when considering the character and pursuits of the people. Undoubtedly under this second head of warlike preparation must come the maintenance of suitable naval stations, in those distant parts of the world to which the armed shipping must follow the peaceful vessels of commerce. The protection of such stations must depend either upon direct military force, as do Gibraltar and Malta, or upon a surrounding friendly population, such as the American colonists once were to England, and, it may be presumed, the Australian colonists now are. Such friendly surroundings and backing, joined to a reasonable military provision, are the best of defences, and when combined with decided preponderance at sea, make a scattered and extensive empire, like that of England, secure; for while it is true that an unexpected attack may cause disaster in some one quarter, the actual superiority of naval power prevents such disaster from being general or irremediable. History has sufficiently proved this. England's naval bases have been in all parts of the world; and her fleets have at once protected them, kept open the communications between them, and relied upon them for shelter.

Colonies attached to the mother-country afford, therefore, the surest means of supporting abroad the sea power of a country. In peace, the influence of the government should be felt in promoting by all means a warmth of attachment and a unity of interest which will make the welfare of one the welfare of all, and the quarrel of one the quarrel of all; and in war, or rather for war, by inducing such measures of organization and defence as shall be felt by all to be a fair distribution of a burden of which each reaps the benefit.

Such colonies the United States has not and is not likely to have. As regards purely military naval stations, the feeling of her people was probably accurately expressed by an historian of the English navy a hundred years ago, speaking then of Gibraltar and Port Mahon. "Military governments," said he, "agree so little with the industry of a trading people, and are in themselves so repugnant to the genius of the British people, that I do not wonder that men of good sense and of all parties have inclined to give up these, as Tangiers was given up." Having therefore no foreign establishments, either colonial or military, the ships of war of the United States, in war, will be like land birds, unable to fly far from their own shores. To provide resting-places for them, where they can coal and repair, would be one of the first duties of a government proposing to itself the development of the power of the nation at sea.

As the practical object of this inquiry is to draw from the lessons of history inferences applicable to one's own country and service, it is proper now to ask how far the conditions of the United States involve serious danger, and call for action on the part of the government, in order to build again her sea power. It will not be too much to say that the action of the government since the Civil War, and up to this day, has been effectively directed solely to what has been called the first link in the chain which makes sea power. Internal development, great production, with the accompanying aim and boast of self-sufficingness, such has been the object, such to some extent the result. In this the government has faithfully reflected the bent of the controlling elements of the country, though it is not always easy to feel that such controlling elements are truly representative, even in a free country. However that may be, there is no doubt that, besides having no colonies, the intermediate link of a peaceful shipping, and the interests involved in it, are now likewise lacking. In short, the United States has only one link of the three.

The circumstances of naval war have changed so much within the last hundred years, that it may be doubted whether such disastrous effects on the one hand, or such brilliant prosperity on the other, as were seen in the wars between England and France, could now recur. In her secure and haughty sway of the seas England imposed a yoke on neutrals which will never again be borne; and the principle that the flag covers the goods is forever secured. The commerce of a belligerent can therefore now be safely carried on in neutral ships, except when contraband of war or to blockaded ports; and as regards the latter, it is also certain that there will be no more paper blockades. Putting aside therefore the question of defending her seaports from capture or contribution, as to which there is practical unanimity in theory and entire indifference in practice, what need has the United States of sea power? Her commerce is even now carried on by others; why should her people desire that which, if possessed, must be defended at great cost? So far as this question is economical, it is outside the scope of this work; but conditions which may entail suffering and loss on the country by war are directly pertinent to it. Granting therefore that the foreign trade of the United States, going and coming, is on board ships which an enemy cannot touch except when bound to a blockaded port, what will constitute an efficient blockade? The present definition is, that it is such as to constitute a manifest danger to a vessel seeking to enter or leave the port. This is evidently very elastic. Many can remember that during the Civil War, after a night attack on the United States fleet off Charleston, the Confederates next morning sent out a steamer with some foreign consuls on board, who so far satisfied themselves that no blockading vessel was in sight that they issued a declaration to that effect. On the strength of this declaration some Southern authorities claimed that the blockade was technically broken, and could not be technically re-established without a new notification. Is it necessary, to constitute a real danger to blockade-runners, that the blockading fleet should be in sight? Half a dozen fast steamers, cruising twenty miles offshore between the New Jersey and Long Island coast, would be a very real danger to ships seeking to go in or out by the principal entrance to New York; and similar positions might effectively blockade Boston, the Delaware, and the Chesapeake. The main body of the blockading fleet, prepared not only to capture merchant-ships but to resist military attempts to break the blockade, need not be within sight, nor in a position known to the shore. The bulk of Nelson's fleet was fifty miles from Cadiz two days before Trafalgar, with a small detachment watching close to the harbor. The allied fleet began to get under way at 7 A.M., and Nelson, even under the conditions of those days, knew it by 9.30. The English fleet at that distance was a very real danger to its enemy. It seems possible, in these days of submarine telegraphs, that the blockading forces in-shore and off-shore, and from one port to another, might be in telegraphic communication with one another along the whole coast of the United States, readily giving mutual support; and if, by some fortunate military combination, one detachment were attacked in force, it could warn the others and retreat upon them. Granting that such a blockade off one port were broken on one day, by fairly driving away the ships maintaining it, the notification of its being re-established could be cabled all over the world the next. To avoid such blockades there must be a military force afloat that will at all times so endanger a blockading fleet that it can by no means keep its place. Then neutral ships, except those laden with contraband of war, can come and go freely, and maintain the commercial relations of the country with the world outside.

It may be urged that, with the extensive sea-coast of the United States, a blockade of the whole line cannot be effectively kept up. No one will more readily concede this than officers who remember how the blockade of the Southern coast alone was maintained. But in the present condition of the navy, and, it may be added, with any additions not exceeding those so far proposed by the government,[8] the attempt to blockade Boston, New York, the Delaware, the Chesapeake, and the Mississippi, in other words, the great centres of export and import, would not entail upon one of the large maritime nations efforts greater than have been made before. England has at the same time blockaded Brest, the Biscay coast, Toulon, and Cadiz, when there were powerful squadrons lying within the harbors. It is true that commerce in neutral ships can then enter other ports of the United States than those named; but what a dislocation of the carrying traffic of the country, what failure of supplies at times,

8. Since the above was written, the secretary of the navy, in his report for 1889, has recommended a fleet which would make such a blockade as here suggested very hazardous.

what inadequate means of transport by rail or water, of dockage, of lighterage, of warehousing, will be involved in such an enforced change of the ports of entry! Will there be no money loss, no suffering, consequent upon this? And when with much pain and expense these evils have been partially remedied, the enemy may be led to stop the new inlets as he did the old. The people of the United States will certainly not starve, but they may suffer grievously. As for supplies which are contraband of war, is there not reason to fear that the United States is not now able to go alone if an emergency should arise?

The question is eminently one in which the influence of the government should make itself felt, to build up for the nation a navy which, if not capable of reaching distant countries, shall at least be able to keep clear the chief approaches to its own. The eyes of the country have for a quarter of a century been turned from the sea; the results of such a policy and of its opposite will be shown in the instance of France and of England. Without asserting a narrow parallelism between the case of the United States and either of these, it may safely be said that it is essential to the welfare of the whole country that the conditions of trade and commerce should remain, as far as possible, unaffected by an external war. In order to do this, the enemy must be kept not only out of our ports, but far away from our coasts.[9]

Can this navy be had without restoring the merchant shipping? It is doubtful. History has proved that such a purely

9. The word "defence" in war involves two ideas, which for the sake of precision in thought should be kept separated in the mind. There is defence pure and simple, which strengthens itself and awaits attack. This may be called passive defence. On the other hand, there is a view of defence which asserts that safety for one's self, the real object of defensive preparation, is best secured by attacking the enemy. In the matter of sea-coast defence, the former method is exemplified by stationary fortifications, submarine mines, and generally all immobile works destined simply to stop an enemy if he tries to enter. The second method comprises all those means and

military sea power can be built up by a despot, as was done by Louis XIV.; but though so fair seeming, experience showed that his navy was like a growth which having no root soon withers away. But in a representative government any military expenditure must have a strongly represented interest behind it, convinced of its necessity. Such an interest in sea power does not exist, cannot exist here without action by the government. How such a merchant shipping should be built up, whether by subsidies or by free trade, by constant administration of tonics or by free movement in the open air, is not a military but an economical question. Even had the United States a great national shipping, it may be doubted whether a sufficient navy would follow; the distance which separates her from other great powers, in one way a protection, is also a snare. The motive, if any there be, which will give the United States a navy, is probably now quickening in the Central American Isthmus. Let us hope it will not come to the birth too late.

Here concludes the general discussion of the principal ele-

---

weapons which do not wait for attack, but go to meet the enemy's fleet, whether it be but for a few miles, or whether to his own shores. Such a defence may seem to be really offensive war, but it is not; it becomes offensive only when its object of attack is changed from the enemy's fleet to the enemy's country. England defended her own coasts and colonies by stationing her fleets off the French ports, to fight the French fleet if it came out. The United States in the Civil War stationed her fleets off the Southern ports, not because she feared for her own, but to break down the Confederacy by isolation from the rest of the world, and ultimately by attacking the ports. The methods were the same; but the purpose in one case was defensive, in the other offensive.

The confusion of the two ideas leads to much unnecessary wrangling as to the proper sphere of army and navy in coast-defence. Passive defences belong to the army; everything that moves in the water to the navy, which has the prerogative of the offensive defence. If seamen are used to garrison forts, they become part of the land forces, as surely as troops, when embarked as part of the complement, become part of the sea forces.

ments which affect, favorably or unfavorably, the growth of sea power in nations. The aim has been, first to consider those elements in their natural tendency for or against, and then to illustrate by particular examples and by the experience of the past. Such discussions, while undoubtedly embracing a wider field, yet fall mainly within the province of strategy, as distinguished from tactics. The considerations and principles which enter into them belong to the unchangeable, or unchanging, order of things, remaining the same, in cause and effect, from age to age. They belong, as it were, to the Order of Nature, of whose stability so much is heard in our day; whereas tactics, using as its instruments the weapons made by man, shares in the change and progress of the race from generation to generation. From time to time the superstructure of tactics has to be altered or wholly torn down; but the old foundations of strategy so far remain, as though laid upon a rock. There will next be examined the general history of Europe and America, with particular reference to the effect exercised upon that history, and upon the welfare of the people, by sea power in its broad sense. From time to time, as occasion offers, the aim will be to recall and reinforce the general teaching, already elicited, by particular illustrations. The general tenor of the study will therefore be strategical, in that broad definition of naval strategy which has before been quoted and accepted: "Naval strategy has for its end to found, support, and increase, as well in peace as in war, the sea power of a country." In the matter of particular battles, while freely admitting that the change of details has made obsolete much of their teaching, the attempt will be made to point out where the application or neglect of true general principles has produced decisive effects; and, other things being equal, those actions will be preferred which, from their association with the names of the most distinguished officers, may be presumed to show how far just tactical ideas obtained in a particular age or a particular service. It will also be desirable, where analogies between ancient and modern weapons appear on the surface, to derive such probable lessons as they offer, without laying undue stress upon the points of resemblance. Finally, it must be remembered that, among all changes, the nature of man remains much the same; the personal equation, though uncertain in quantity and quality in the particular instance, is sure always to be found.

CHAPTER III

FOUNDATIONS AND PRINCIPLES

THE SEARCH FOR and establishment of leading principles—always few—around which considerations of detail group themselves, will tend to reduce confusion of impression to simplicity and directness of thought, with consequent facility of comprehension. It must be noted likewise that while steam has facilitated all naval movements, whether strategical or tactical, it has also brought in the element of communications to an extent which did not before exist. The communications are, perhaps, the most controlling feature of land strategy; and the dependence of steam ships upon renewing their limited supply of coal, contrasted with the independence of sailing ships as to the supply of their power of motion, is exactly equivalent to the dependence of an army upon its communications. It may be noted, too, that, taking one day with another, the wind in the long run would average the same for each of two opponents, so that in the days of sail there would be less of the inequality which results from the tenure of coaling stations, or from national nearness to the seat of war. Coal will last a little longer, perhaps, than the supplies an army can carry with it on a hurried march, but the anxiety about it is of the same character; and in the last

Chapter VI, *Naval Strategy* (Boston, 1911), 118–31.

analysis it is food and coal, not legs and engines, which are the motive powers on either element.

The days when fleets lay becalmed are gone, it is true; but gone also are the days when, with four or five months of food and water below, they were ready to follow the enemy to the other side of the world without stopping. Nelson, in 1803–1805, had always on board three months' provisions and water, and aimed to have five months'; that is, to be independent of communications for nearly five months. If it is sought to lessen the strategic difficulty by carrying more coal, there is introduced the tactical drawback of greater draught, with consequent slower speed and more sluggish handling; or, if tonnage is not increased, then armor and guns are sacrificed, a still more important consideration. The experience of Admiral Rozhestvensky in this matter is recent and instructive. His difficulties of supply, and chiefly of coal, are known; the most striking consequence is the inconsiderate manner in which, without necessity, he stuffed his vessels with coal for the last run of barely a thousand miles. That he did this can be attributed reasonably only to the impression produced upon his mind by his coaling difficulties, for the evident consequence of this injudicious action was to put his ships in bad condition for a battle which he knew was almost inevitable.

Both the power and the difficulties due to steam call for a more comprehensive and systematic treatment of the art of war at sea, and for the establishment of definite principles upon which it reposes. To do this is simply a particular instance of the one object for which the War College exists. As the principles of the art of war are few, while embracing many features, so the principle of the War College is one; namely, the study of the art of war and the exposition of its principles. Like the body, it has but one backbone though many ribs. When these principles have been more or less successfully defined, the way is open to a clearer comprehension of naval history, a more accurate perception of the causes of success or failure in naval campaigns. Study of these, superimposed upon an adequate grasp of principles, contributes to the naval strategist the precise gain which the practice of a profession gives to a man—a lawyer, for instance—who has already mastered the principles. Extensive study of cases gives firmer grasp, deeper understanding, wider views, increased aptitude and quickness to apprehend the critical features in any suit, as distinct from details of less relative importance.

When I was a midshipman, a very accomplished officer, the late Admiral Goldsborough, told me of his bewilderment in listening to the arguments of eminent lawyers in a difficult suit. Later in the day, meeting the judge who presided, he said to him, "Upon my word, I don't understand how you can see your way through such a maze of plausibilities as were presented by the two sides." The judge replied, "There are in such contentions a very few, perhaps only one or two, really decisive considerations of fact or principle. Keeping those firmly in mind, much of the argument sheds off, as irrelevant, or immaterial, and judgment is therefore easy." This is the advantage of the habit of mind bred by study, when principles are understood. Such decisive considerations correspond essentially to the leading feature, or features, which constitute "the key" of a military situation.[1] A mass of confusing incidents group themselves around certain decisive considerations, by holding which firmly you not only understand more easily the determining factors in a particular case, but are fit-

1. Clausewitz pokes some mild fun at the expression "the key" of a military theater, or situation; which probably does come too easily to the lips, or to the pen, as if in itself conveying an encyclopædia of explanation. The analogies of the use of the word in other relations, however, justify its application to military conditions. In the judgment of the writer, the use of it has the special advantage of conducing to sustain the desirable impression that in most military situations, or problems, there is some one leading feature, so far primary, that, amid many important details, it affords a central idea upon which concentration of purpose and dispositions may fasten, and so obtain unity of design.

ting yourself more and more to judge any military case put before you; and that, too, with the rapidity for which military urgency often calls.

Here is seen the value of land warfare to the naval student. In the first place, land warfare has a much more extensive narrative development, because there has been very much more land fighting than sea; and also, perhaps because of this larger amount of material, much more effort has been made to elicit the underlying principles by formal analysis. Further, with the going of uncertainty and the coming of certainty into the motive power, a chief distinction between the movements of fleets and armies has disappeared. Unless, therefore, one is prepared to discard as useless what our predecessors have learned, it is in the study of the best military writers that we shall find the most ample foundations on which to build the new structure. Not attempting the vain, because useless, labor of starting on unbroken ground, we will accept what is already done as clear gain, and build. No doubt—and no fear—but we shall find differences enough; no one will mistake the new house for the old when it is finished; yet the two will have a strong resemblance, and the most marked contrasts will but bring out more clearly than ever the strong features common to both.

The definitions of strategy, as usually given, confine the application of the word to military combinations, which embrace one or more fields of operations, either wholly distinct or mutually dependent, but always regarded as actual or immediate scenes of war. However this may be on shore, a French writer is unquestionably right in pointing out that such a definition is too narrow for naval strategy.

> "This," he says, "differs from military strategy, in that it is as necessary in peace as in war. Indeed, in peace it may gain its most decisive victories by occupying in a country, either by purchase or treaty, excellent positions which would perhaps

hardly be got in war. It learns to profit by all opportunities of settling on some chosen point of a coast, and to render definitive an occupation which at first was only transient."

This particular differentiation of naval strategy is due to the unsettled or politically weak conditions of the regions to which navies give access, which armies can reach only by means of navies, and in which the operations of an army, if attempted, depend upon control of the sea. If a nation wishes to exert political influence in such unsettled regions it must possess bases suitably situated; and the needs of commerce in peace times often dictate the necessity of such possessions, which are acquired, as the French writer says, when opportunity offers.

In Europe, the great armies now prevent such acquisitions, except at the cost of war; although it is perhaps a little difficult to maintain this statement in the face of the recent annexation of Bosnia and Herzegovina. In truth, however, southeastern Europe, owing to the weakness of Turkey, brings to the back door of Europe just that sort of condition which for the most part is to be found only in the remoter regions which navies reach; while the political upset in Turkey gave the opportunity, and the pretext, for Austria to act,—to consolidate her power in a strategic position, which, to say the least, advances her towards the Ægean, a goal desirable for her commercial future. Thus, also, passing over more ancient historical instances, England within ten years of peace occupied Cyprus and Egypt under terms and conditions on their face transient; but which in the former case have led to a formal cession, and in the latter, after over a quarter of a century, have not yet ended in an evacuation of the country. She there still holds the possession which is nine points of the law, despite the long continued, but at last apparently appeased, discontent of France and Russia.

Similarly, in later years, France has possessed herself of the territory of Tunis and of its port Bizerta, the possibilities of which as a naval station are highly spoken of, and appear to be superior to those of Algiers in important hydrographical particulars, as well as in nearness to a narrow part of the Mediterranean; in closeness, that is, to the necessary line of communication between the Straits of Gibraltar and the Suez Canal, the critical link in the European route to the Far East, to India, and to Australia. Again, there is the German position of Kiao-Chau, concerning the concession of which by China the Chancellor of the German Empire said at the time that the need of a base in the Far East, for both commercial and political—that is, naval—reasons, had long been foreseen. Consequently upon opportune occasion it had been acquired by pressure upon China. The Caroline and other Pacific Islands purchased by Germany since these lectures were first written are illustrations of the truth, stated by the French writer quoted, that "Naval Strategy has for its end to found, support, and increase, *as well in peace as in war,* the sea power of a country." I doubt, indeed, whether the same is not true of land strategy; but the positions in which it is interested—the scenes of land warfare—are so well understood, and so firmly held by long prescriptive right, that they cannot ordinarily be transferred, except at the cost of war. The diplomatist, as a rule, only affixes the seal of treaty to the work done by the successful soldier. It is not so with a large proportion of strategic points upon the sea. The above positions have all been acquired in peace, and without hostilities. The same is true of the acquisition of the Hawaiian Islands by the United States, accomplished long after the writing of these lectures. Such possessions are obtained so often without actual war, because the first owners on account of weakness are not able to make the resistance which constitutes war; or, for the same reason of weakness, feel the need of political connection with a powerful naval state.

Closely associated with this point of view, which depends

upon the usual remoteness of the positions thus acquired from the country acquiring them, is the very large geographical scale upon which naval operations are conducted as compared with those on land. This was an aspect which struck the late General Sherman, when he did me the favor to read the original draft of these lectures. Bases coincident with the whole seacoast of a country, lines of communication hundreds of miles long, leading to objectives equally remote, movements at the rate of hundreds of miles in a day, impressed the imagination which had conceived and effected the noted March to the Sea.

Another illustration of naval strategy in time of peace, which also depends in large measure upon the great distances which separate the strategic centers of interest,—centers, for example, such as those of the Atlantic and Pacific coasts of the United States, or those of Great Britain in the Narrow Seas and in the Mediterranean,—is to be seen in the changed disposition of navies at the present time. It would be interesting to estimate how much this is due to circumstances, to changes in international conditions, and how much to the greater attention to and comprehension of the principles and requirements of strategy, now to be found in naval officers, as compared with the placid acquiescence of former generations in routine traditions. I think it would be safe to say in this connection that the present recognition of the necessity for concentration is an advance due to study, to intellectual appreciation of a principle and of the military ineptness and danger of the former method of distributing the force of a nation in many quarters during peace; but that the particular methods in which this appreciation has shown itself are the result of international conditions. As an instance may be cited the present concentration of the British fleet in home waters. This is an immediate reflection of German naval development. A corollary to this change in the distribution of the fleet is the enhanced importance of the Chatham dockyard and the initiation of a new base at Rosyth. Both are illustrations of strategic positions established or developed in peace.

Another instance, of more value for an analysis, is the concentration of the United States battle-fleet in one command and one body. This illustrates the effect of the simple principle upon the minds of those who direct the navy, and also has a particular indicative value; for international relations have not as yet compelled that concentration to be localized, either in the Atlantic or Pacific, as that of Great Britain has been in home waters. The concentration is due to a simple recognition of principle, not to pressure of circumstances. It is known in the navy, however, that the recognition was first made in the process of war games at this College. That this concentration at present is in the Atlantic is merely the continuance of a tradition that our chief danger is from Europe, as for a long time was the case. This may be true now, or it may not; circumstances, that is, the developments of international relations, will determine from time to time the place of concentration, as it has for Great Britain. In connection with this line of thought, let it be noted that in the round-the-world cruise of the battle-fleet, a conspicuous matter for observation was the disappearance of the small squadrons and scattered vessels which once testified the general naval policy of governments.

The necessity for such sustained naval concentrations depends again upon the characteristic which above all differentiates naval strategy from that upon land. This characteristic is the mobility of navies as compared with armies, the outcome of the very different surfaces over which they respectively move. A properly disposed fleet is capable of movement to a required strategic position with a rapidity to which nothing on land compares. This necessitates a corresponding preparation on the other side, which at the least must be ready to get there equally rapidly and equally concentrated. All this is mobilization; a process common to land and sea, but differing both in the scale and in the rapidity with which it can be conducted. At sea, for navies, the process also is simple; which again means that it can be rapid. Complication means loss of time. For these reasons, while the disposition of armies in peace must be maintained with direct reference to war, the difficulty of mobilization for the other party permits a dispersion of the forces on land which is impolitic in naval dispositions. In the mobilization of a land force, concentration, militarily understood, is the prime object, as it is with navies; but it is the second step, that is, it follows the local activities which mobilize the several corps. With navies it should be less the first step than the condition at the instant war breaks out, however unexpected. Then again the impedimenta, the train, which constitutes so large a factor in military movements, exists for navies only in a very modified degree; and the train possesses substantially the same mobility as the battleships themselves, because the open field of the sea offers wider facilities than roads can do. All these advantages in mobility mean rapidity in time; and this reduction in the scale of time required for movement means expansion in the scale of distance that can be covered, in order to overpower a dispersed or an unwary enemy. Thus when the Japanese torpedo vessels surprised the unready Russian fleet before Port Arthur, they opened hostilities some hundreds of miles from their point of departure.

"The possession of the strategic points," says the Archduke Charles, "decides the success of the operations of war." This Napoleon also expressed in the words, "War is a business of positions." It is necessary, however, to guard against a mistake so common that it seems almost to be a permanent bias of the human mind in naval matters. It is one that has come home to myself gradually and forcibly throughout my reading; a result which illustrates aptly what I have just said of the gain by reading widely after principles are understood. I knew long ago, and quoted in these lectures, Jomini's assertion that it is possible to hold too many strategic points; but it is only by subsequent reading that I have come to appreciate how common is the opinion that the holding of each additional port adds to naval strength. Naval strength involves, unquestionably, the possession of strategic points, but its greatest constituent is the mobile navy. If having many ports tempts you to scatter your force among them, they are worse than useless. To this is to be added another remark, also due to Jomini, that if you cannot hope to control the whole field, it is an advantage to hold such points as give you control of the greater part of it. The farther toward an enemy you advance your tenable position by the acquisition of strategic points, or by the positions occupied in force by army or navy, the better; provided, in so doing, you do not so lengthen your lines of communication as to endanger your forces in the advanced positions.

An exceptionally strong illustration of the benefit of such advanced position is afforded by the Island of Cuba and the effect exercised upon the control of the Gulf of Mexico, according as a position in that island may be held or not by the United States. While Cuba was Spanish, the United States had to depend upon Pensacola and the Mississippi as points upon which to base naval operations. If, in such conditions, war arose with a European state, Cuba being neutral, the enemy venturing his battle-fleet into the Gulf of Mexico would not thereby expose his rear or his communications to attack in force to the same extent that he would now with the United States cruisers based upon Guantanamo, duly fortified. Between opponents of equal force this advanced position gives a decided advantage to the occupant by the facility it affords to molest and interrupt the supplies, and especially the coal supplies, of a hostile fleet attempting to maintain itself within the Gulf, or advanced in the Caribbean towards the Isthmus. As regards the Gulf coast alone, Key West to some extent would fulfil the office of Guantanamo. The two together are a better defence for our Gulf region as a whole than localized land defences at particular points of the region would be. As regards influence over the Canal Zone, the superiority in situation of Guantanamo over Key West is obvious. The deterrent effect of such positions upon a fleet does not apply to the same degree to single fast cruisers or small squadrons, because the loss of a few of them can be risked for the sake of annoying an enemy.

The supreme naval instance of an advanced position in former times was the British blockade of French ports, by which the safety of British commerce was assured and the invasion in force of the British Islands prevented. A closely analogous disposition is the present concentration of the British battlefleet in the home waters, having in view, as is well understood, immediate effective concentration against Germany in the North Sea. In case of war, whatever particular measures may be then adopted, the presence there of a fleet decisively superior to the German covers effectively all British lines of communication from the Atlantic; that is, practically with the entire world, with a possible exception of the Baltic countries. The same disposition intercepts all German sea communications except with the Baltic. It also covers the British Islands against an invasion in force.

From these instances the general reason for taking up such an advanced position is obvious. Behind your fleets, thus resting on secure positions and closely knit to the home country by well-guarded communications, the operations of commerce, transport, and supply can go on freely. Into such a sea the enemy cannot venture in force about equal to your own,—Germany, in the instance just cited, into the Atlantic, or an enemy of the United States into the Gulf,—because in the very act of venturing he exposes his communications, and, in case of reverse, he is too far away from his home ports. Cuba thus covers the Gulf of Mexico, but would not have an equal material effect upon operations against the North Atlantic Coast. The British blockades of a century ago, on the contrary, being pushed right up to the French shores, covered the entire ocean and all approaches to the British Islands, because so far advanced. In virtue of that advance, while maintained, they conferred upon the home country perfect security from invasion with substantial immunity to the commerce of the United Kingdom, the loss being less than three per cent per annum.

To-day, the British Islands by their geographical situation alone, as towards Germany, themselves occupy an advanced position; their control over the North Sea resembles closely that of Cuba over the Gulf of Mexico, and their defensive value to the communications of the country are the same. Even German cruisers,—commerce destroyers,—to reach the British commercial communications, must run the gauntlet of the North Sea, and act with diminished coal supply far from their bases of operation. The rear and its communications cannot, we know, be protected wholly from commerce destroyers in their attacks either upon supply ships or commerce. Such raids on the flanks and rear of an army were frequent in the American War of Secession. They can only be checked, not wholly prevented, by light bodies, or by cruisers similar to those who make them.

> "Good partisan troops," says Jomini, whose experiences antedated the American War of Secession by half a century, "will always disturb convoys, whatever be the direction of the roads, even were that direction a perpendicular from the center of the base to the center of the front of operations—the case in which they are least open to the attacks of an enemy."

Such injuries, however, are not usually to be confounded with the cutting, or even threatening, the communications. They are the slight wounds of a campaign, not mortal blows; vexatious, not serious. It is a very different matter to have a powerful fleet in a strong port close to the communications.

Raiding operations against commerce, or against an enemy's communications, may proceed from remote colonial positions. In former wars the French West India Islands, Martinique and Guadeloupe, thus served as bases for French cruisers against British commerce and supply vessels. Provision against these raids did not then, and cannot now, depend directly upon the distant home country. They must be met by local dispositions. Such positions themselves illustrate particular cases of advanced positions, exercising a specific, if limited, control. For instance, German Southwest Africa, as far as situation goes, has facilities for molesting British intercourse with the Cape of Good Hope, or beyond by that route. To meet such a condition provision likewise must be local. The effect of the British concentration in the North Sea is in such cases indirect, though real. It imposes the question how far such detachments, made before war, would be consistent with the general scheme of German North Sea operations; with the further problem how far the detachments could be sustained in efficiency, in face of the difficulty of passing supplies through the lines of British cruisers in the North Sea and Channel.

A distinct and great reinforcement to the effect of a line of advanced positions is that it be continuous by land, and extensive. Thus the ports of Cuba have value additional to their individual advantages, from the fact that they not only are connected by land, but that this land barrier is nine hundred miles long, a troublesome obstacle to an enemy. In the same way, the effect of the British Islands upon North Sea commerce is increased by the continuousness of the land from the Straits of Dover to the north end of Scotland.

The determination, therefore, of the strategic points of a maritime area, such as the Gulf and Caribbean, or as the Pacific, the two seas in which the United States is most critically interested, must be followed by a selection from among them, first, of those which have the most decisive effect upon the control of the theater of war; secondly, of those which represent the most advanced position which the United States, in case war unhappily arose, could occupy firmly, linked to it by intermediate positions or lines, such that the whole would form a well-knit, compact system from which she could not be dislodged by any but a greatly superior force.

CHAPTER IV

POSITION, STRENGTH, RESOURCES

1. STRATEGIC POSITIONS

THE STRATEGIC VALUE of any place depends upon three principal conditions:

1. Its position, or more exactly its situation. A place may have great strength, but be so situated with regard to the strategic lines as not to be worth occupying.

2. Its military strength, offensive and defensive. A place may be well situated and have large resources and yet possess little strategic value, because weak. It may, on the other hand, while not naturally strong, be given artificial strength for defense. The word "fortify" means simply to make strong.

3. The resources, of the place itself and of the surrounding country. It is needless to explain the advantages of copious resource or the disadvantages of the reverse. A conspicuous example of a place strong both for offense and defense, and admirably situated, yet without natural resources, is Gibraltar. The maintenance of this advanced post of Great Britain depended in the past wholly upon her control of the sea. Resources that are wanting naturally may be supplied artificially, and to a greater extent now than formerly. Malta and Minorca illustrate the same truth but to a less degree, and

Chapter VII, *Naval Strategy* (Boston, 1911), 132–63.

generally, in sea strategic points, the smaller the surrounding friendly territory, the fewer the resources and the less the strength. In 1798–1800 the French garrison at Valetta was cut off from the resources of the island of Malta by the revolt of the islanders, supported by the British; and being rigidly blockaded by sea, its resistance was ended by exhaustion. From these considerations it follows that, other things being equal, a small island is of less strategic value than a large one; and a point like Key West, at the end of a long narrow peninsula of restricted access, is in so far inferior to Pensacola, and would be to Havana or Cienfuegos if Cuba were a thriving country.

As an illustration of the advantage of a large island over a small, or over several small ones, I will read you the opinion of the well-known Admirable Rodney, found in an official memorandum of the period of the War of American Independence. Rodney had had a very long experience of the West Indies, both in peace and war.

> Porto Rico, in the hands of Great Britain, will be of infinite consequence, and of more value than all the Caribbee Islands united—will be easily defended, and with less expense than those islands; the defense of which divides the forces, and renders them an easier conquest to an active enemy: but this island will be such a check to both France and Spain, as will make their island of St. Domingo be in perpetual danger, and, in the hands of Great Britain, enable her to cut off all supplies from Europe bound to St. Domingo, Mexico, Cuba, or the Spanish Main; and, if peopled with British subjects, afford a speedy succour to Jamaica; and, when cultivated, employ more ships and seamen than all the Windward Islands united.

In this you have an example of the material which, as I have said before, naval history furnishes in abundance to the student of the art of war. All the advantages of a strategic point are here noted, though not quite in the orderly, systematic manner at which a treatise on the art of war would aim: Situation, relatively to Jamaica, Santo Domingo and other Spanish possessions; defensive strength, due to concentration, as compared with the dispersion of Lesser Antilles; offensive strength as against the communications of Spain with her colonies; and resources of numerous British subjects with their occupations, as well as of British ships and seamen.

Where all three conditions, situation, intrinsic strength, and abundant resources, are found in the same place, it becomes of great consequence strategically and may be of the very first importance, though not always. For it must be remarked that there are other considerations, lesser in the purely military point of view, which enhance the consequence of a seaport even strategically; such as its being a great mart of trade, a blow to which would cripple the prosperity of the country; or the capital, the fall of which has a political effect additional to its importance otherwise.

Of the three principal conditions, the first, situation, is the most indispensable; because strength and resources can be artificially supplied or increased, but it passes the power of man to change the situation of a port which lies outside the limits of strategic effect.

Generally, value of situation depends upon nearness to a sea route; to those lines of trade which, when drawn upon the ocean common, are as imaginary as the parallels of the chart, yet as really and usefully exist. If the position be on two routes at the same time, that is, near the crossing, the value is enhanced. A cross-roads is essentially a central position, facilitating action in as many directions as there are roads. Those familiar with works on the art of land war will recognize the analogies. The value becomes yet more marked if, by the lay of the land, the road to be followed becomes very narrow; as at the Straits of Gibraltar, the English Channel, and in a less degree the Florida Strait. Perhaps narrowing should be applied to every inlet of the sea, by which trade enters into and is distributed over a great extent of country; such as the mouth of the Mississippi, of the Dutch and German rivers, New York harbor, etc. As regards the sea, however, harbors or the mouths of rivers are usually *termini* or *entrepôts,* at which goods are transshipped before going farther. If the road be narrowed to a mere canal, or to the mouth of a river, the point to which vessels must come is reduced almost to the geometrical definition of a point and nearby positions have great command. Suez presents this condition now, and Panama soon will.

Analogously, positions in narrow seas are more important than those in the great ocean, because it is less possible to avoid them by a circuit. If these seas are not merely the ends—"*termini*"—of travel but "highways," parts of a continuous route; that is, if commerce not only comes to them but passes through to other fields beyond, the number of passing ships is increased and thereby the strategic value of the controlling points. It may, perhaps, be well to illustrate here, by the instance of the Mediterranean, the meaning I attach to the words "*termini*" and "highways." Before the cutting of the Suez Canal, the Levant and Isthmus were *termini.* Ships could not pass; nor goods, except by transshipment. Since the canal, the Levant has become a point on a highway and its sea is a highway of trade, not a *terminus* only. The same remarks apply, of course, to the American Isthmus and any future canal there. If Bermuda be compared with Gibraltar, or even with Malta, as to position only, the advantage of these will be seen at once and the argument concerning narrow seas illustrated; for shipping must pass close by them, while Bermuda, advantageous though it is, regarded as a depot, and favorably situated for offensive operations against usual trade routes, may be avoided by a circuit, involving inconvenience and delay, but still possible.

A radical difference underlying the conditions of land and sea strategy is to be found in the fact that the land is by nature full of obstacles, and removing or overcoming of which by men's hands opens communications or roads. By nature, the land is almost all obstacle, the sea almost all open plain. The roads which can be followed by an army are therefore of limited number, and are generally known, as well as their respective advantages; whereas at sea the paths by which a ship can pass from one point to another are innumerable, especially if a steamer, content to make a circuit. The condition of winds, currents, etc., certainly do combine with shortness of distance to tie ships down to certain general lines, but within these lines there is great scope for ingenuity in dodging the search of an enemy. Thus Rodney, in a despatch to the Admiralty concerning a homeward-bound convoy from the West Indies, states that, instead of going direct, it will proceed so as to reach the latitude of the English Channel at least six hundred miles west of it, and thence steer due east, thus deceiving the enemy as to its position, as well as enabling the Admiralty with certainty to reinforce the protecting ships. On a later occasion he wrote, "I have given the commanding officer the strictest orders on no account to attempt the Channel, but to gain the latitude of Cape Clear at least nine hundred miles west of that cape and thence proceed." So Napoleon in one of his condensed phrases said that the determining elements in naval operations were "*Fausses routes et moments perdus.*"

A very pertinent historical instance was in the pursuit of Bonaparte's Egyptian expedition by Nelson in 1798. The French commander-in-chief, after leaving Malta, laid his course first for Crete, instead of towards Egypt. Nelson, satisfied as to the enemy's destination, naturally and properly pushed direct for Egypt. Unluckily, he had not a single frigate for lookout service. His track consequently diverged from that of the French, and he lost them; the wake of the two fleets actually crossed on the same night, but a light haze hid them from each other. Such conditions made it necessary for Great Britain during the great wars to keep a close lookout, if not a blockade, at the entrance of the French harbors, which thus became strategic points; for if the fleet within once got away and was lost to sight, nothing was left to the British commander but to reason out as well as he could their probable line of action.

An interesting illustration of the essential similarity of conditions, under all the qualifications introduced by modern development, is to be found in comparing the perplexities of Admiral Togo in 1905 with those of Nelson in 1798. Nelson did not *know* whither the French were bound; he depended upon inference, deduced from indications and from existing political conditions. Togo did not *know* what the Russians might attempt, whether fight or flee, though their ultimate destination could be only Vladivostok; but as to the route they would take he had to depend upon inference, in which known weather conditions played a large part. Both admirals calculated rightly; but both underwent intervals of anxious suspense, because of want of information. "Even Admiral Togo," wrote one of his staff, "certain as he had felt that the enemy must go by way of Tsushima, began to be anxious as the days of their expected arrival passed without their appearing." Allowing for what wireless does, it may be said without exaggeration that Togo did not learn where the enemy was and what he was doing any sooner than Nelson did; that is, till he was seen by the Japanese scouts. He had lost touch, or rather never had gained it other than by means of common report, unofficially; not by his own vessels. Like Nelson at the Nile, Togo had no certainty before the enemy was seen; in the one case from a masthead of a battleship, in the other by a lookout vessel only a hundred miles distant from the flagship. Both dilemmas arose from failure to watch the enemy in his port of departure or at some unavoidable point of passage. Whether any blame attaches for this failure is another question; the inconvenience was due to scouting not having been pushed far enough forward.

For the reason that the open ocean offers such large opportunity for avoiding a position recognized as dangerous, first-rate strategic points will be fewer within a given area on sea than on land,—a truth which naturally heightens the strategic value of such as do exist. For instance, Hawaii, in the general scheme of the Pacific, is a strategic point of singular importance. It is a great center of movement, an invaluable half-way house, an advanced position of great natural power of offense as a base of operations and for supply and repair; but in the control of commerce its effect is lessened by the wide sweep open to vessels wishing to avoid it. On the other hand, possession gives it defensive value additional to offensive, by excluding an enemy from using it, whether for war or for commerce. The sea, indeed, realizes a supposed case of the Archduke Charles. He says,

> In open countries, which are everywhere practicable, and in which the enemy can move without obstacle in every direction, there are either no strategic points or there are but few; on the contrary, many are to be met in broken countries, where nature has irrevocably traced the roads which must be followed.

As a ship goes from Europe to Central America she passes first through a wholly open country until she reaches the West Indies; there she enters one that is broken, and abounding in strategic points of greater or less value.

The *amount of trade* that passes enters into the question as well as the *nearness of the port* to the route. Whatever affects either affects the value of the position. It is the immense increase of German industry, commerce, and shipping that has made Great Britain, by the strategic position of the British Islands, the menacing object she has become in German eyes. The growth of German trade, combined with the strategic position of Great Britain, has revolutionized the international relations of Europe. A similar new commercial condition, the

Panama Canal, will change the strategic value of nearly every port in the Caribbean and of many in the Pacific, because of the consequent increase of trade passing that way. Imagine Suez closed again forever, and consider the twofold effect,—upon the Cape of Good Hope ports and upon those of the Mediterranean. Of this we have historical demonstration in the effect upon the fortunes of Venice and Genoa from the discovery of the Cape route. Sea power primarily depends upon commerce, which follows the most advantageous roads; military control follows upon trade for its furtherance and pro-protection. Except as a system of highways joining country to country, the sea is an unfruitful possession. The sea, or water, is the great medium of circulation established by nature, just as money has been evolved by man for the exchanges of products. Change the flow of either in direction or amount, and you modify the political and industrial relations of mankind.

In general, however, it will be found that by sea, as by land, useful strategic points will be where highways pass, and especially where they cross or converge; above all, where obstacles force parallel roads to converge and use a single defile, such as a bridge. It may be remarked here that while the ocean is easier and has, generally, fewer obstacles than the land, yet the obstacles are more truly impassable. Ships cannot force their way over or through obstacles, but must pass round them—turn them. Historical feats, such as those of Napoleon crossing the Little St. Bernard, Macdonald the Splügen, and the Russians in 1877 the Balkans, seem to show that nothing is impassable to infantry; but modern ships are not to be dragged over dry land, like the ancient galleys. Hence, while on land the defenses of what may seem to be the only practicable road may be unexpectedly turned by an army, as the Persians by a mountain path reached the rear of the Greeks at Thermopylæ, assurance can be felt that ships can follow only known tracks. Where there are many of these, as, for instance, the passages between the Windward West India Islands, the situation-value of the ports at each passage is proportionately diminished.

Consider, for instance, the enormous effect upon the value of Port Royal, Martinique, and Port Castries in Santa Lucia—already good strategic points—if a continuous line of land extended from the east end of Haiti, through the Windward Islands, to South America, broken only at the passage between the two islands named. Their influence then would be almost identical with that of Gibraltar. As it is, they are in a class with Hawaii and Bermuda; and lower in the class, because their positions, though excellent, are less unique than either. They have rivals within their respective areas, while the two others have not. There can be no question that, whatever the intrinsic military strength of the ports in the Windward Islands, their situation-value is seriously lowered by the fact that an enemy's shipping or supply vessels bound to the Isthmus can, by a circuit, avoid passing near. Jamaica cannot be equally avoided, still less the Chiriqui Lagoon, least of all Colon when the Panama Canal becomes a fact. It is possible, for instance, that in war between Great Britain and France a ship bound to the Isthmus, wishing to avoid passing near Santa Lucia, could go through the Anegada or Mona Passage; and, in fact, such evasions were often successfully resorted to by the French, to avoid Rodney's lying in wait.

2. MILITARY STRENGTH

We come now to the second element in the strategic value of any position, namely, its military strength, offensive or defensive.

It is possible to imagine a point very well placed yet practically indefensible, because the cost of defensive works would be greater than the worth of the place when fortified. A much stronger site, although somewhat further off, would throw such a position out of consideration.

There are several elements, advantageous or disadvantageous, which enter into the characteristics making a port strong or weak, but they will all be found to range themselves under the two heads of defensive and offensive strength.

I. Defensive strength

The defense of seaports, as distinguished from the offensive use made of them, ranges under two heads: 1. Defense against attack from the sea; that is, by ships. 2. Defense against attack from the land; that is, by troops which in the absence of resistance may have landed at some near point of the coast and come up in the rear of the fortress.

As offensive efforts made from a fortified seaport, to facilitate which it has been fortified, are always by ships toward the sea, the sea may properly be spoken of as the front of such a port, while the land side is the rear.

The recent siege of Port Arthur has illustrated the propositions just advanced. Port Arthur was defended against naval attack and against land attack, in front and in rear, and attack was made from both quarters. The siege illustrated also another proposition, made in the original draft of these lectures, that the defense of ports, in the narrow sense of the word "defense," belongs chiefly to the army. The Russian navy contributed little to the defense. If it had been in better moral and material condition, and efficiently used, it might have contributed very materially to the endurance of the port by offensive operations; by sorties, by harassing the enemy. Endurance is a principal element of defensive strength in any general strategic scheme. The great gain of defense, in the restricted sense of the word, is delay. The defense of Port Arthur gained time for the Russians; had the defense been more obstinate, more time still would have been gained. As it was, ample opportunity was obtained for the Baltic fleet to arrive; and no one can tell how far the delay contributed to prolong the land campaign, which, as it was, left Japan with the most of her work still before her, if Russia stood firm.

An illustration of the character of operations by which a navy contributes to defense, to delay, was afforded in the same war by the capture at sea of a Japanese transport carrying a large part of the siege guns for the siege of Port Arthur. This sensibly prolonged the siege. It was an attack upon the communications of the besiegers. Attacks of that character, besides the actual injury inflicted, necessitate to the enemy an elaboration of precautions which sensibly protracts the issue. Such action, however, though defensive in result, is not so in method. It therefore is called, properly enough, an offensive-defensive, and is absolutely essential to any scheme of defense. Napoleon said that no position can be permanently maintained if dependent upon defense only; if not prepared for offensive measures, or if it fails to use them. The enemy must be disturbed or he will succeed. At one time in the history of war this truth was so clearly apprehended, and the conditions of passive resistance so thoroughly appreciated, that the endurance of a besieged fortress could be calculated almost as exactly as a mathematical solution; that is, granting no attempt at relief. In a properly coördinated system of coast defense this counter-action, molestation, the offensive-defensive, belongs to the navy.

Coast defense in the restricted sense, when action is limited to repelling an immediate attack, is the part of the army chiefly; hence the scheme of preparations for such defense also belongs primarily to the army. That being the case, it is not for naval officers to distribute the preparations among the branches of the military service; but it is permissible to note that the duty of planning fortifications and superintending their construction is by accepted tradition assigned to military engineers.

It should be noted also that such tactical considerations as the extent of the outer lines necessary to cover the landward defensive works of a place, and the consequent numbers of the garrison required to insure their maintenance, are questions of expert military knowledge. It follows, and will be still more evident as the naval requirements develop in the ensuing treatment, that sound decision in the selection of naval stations at home and abroad is for combined military and naval consultation. Indeed, every question and every preparation touching seacoast operations present this feature of combination between army and navy, working to a common end.

In all such coöperations there will be found conflicting conditions, as there will in most plans of campaign and in positions taken for battle,—strong here, weak there. War in all its aspects offers a continual choice of difficulties and advantages. It is in reconciliation effected among these as far as possible, in allowance of due predominance to the most important, in disregard of difficulties where practicable, that the art of the commander consists. The one most demoralizing attitude is that which demands exemption from risks, or is daunted unduly by them.

The siege of Port Arthur illustrated another truth, which will be found of general application; namely, that coast fortresses are in greater danger of capture by land attacks than by those from the sea. Santiago showed the same, imperfect as were its sea defenses. The reason is obvious: no vessel, no construction resting upon the water, can bear the same weight of ordnance and the same armor that a land work can. To this inferiority modern warfare has added the additional danger to ships of the submarine mine, the effects of which upon the movements possible to vessels were so often and so strikingly illustrated in the war between Japan and Russia. No similar equal danger exists for land fortifications. In brief, ships are unequally matched against forts, in the particular sphere of forts; just as cavalry and infantry are not equal, either to the other, in the other's proper sphere. A ship can no more stand up against a fort costing the same money, than the fort could run a race with a ship. The quality of the one is ponderousness, enabling great passive strength; that of the other is mobility.

Countries which are entirely surrounded by water, or whose land frontiers are bordered by communities of much less military strength, as is the case with Great Britain and the United States, may easily fall into the error of defending ports only against attack from the sea. For ports which are commercial only, not essential to naval activities, this must answer; for there is a limit to the money that can be spent upon coast fortifications. But any scheme of naval activity rests upon bases, as do all military operations. Bases are the indispensable foundations upon which the superstructure of offense is raised. Important naval stations, therefore, should be secured against attack by land as well as by sea. To illustrate this fact was the aim of General Wood in recent conjoint operations about Boston, and the ease with which the city fell shows the need of defense by land. Purely commercial cities are defended sufficiently by the condition that a large hostile expedition will be employed only in securing an adequate result, a decided military gain, such as the destruction of a great naval base; while a small landing force, though it conceivably might capture a commercial port, can do so only by a surprise, which in effect is a mere raid, liable to interception, and in any event productive of no decided military advantage.

In the English maneuvers of August, 1888, it was found, as might have been predicted, that it is impossible for a blockading force to prevent the escape of single ships. When such had escaped, there was shown, first, what has already been said as to the perplexity of the blockade as to the direction taken; and second, the futility of depending upon the navy alone for the defense of seaports. The escaped cruisers appeared before half a dozen English ports, which in the absence of fortifications had at once to admit their powerlessness and to pay ransom.

The ease of running a blockade has been very much qualified since 1888 by the development given to stationary submarine mines placed by the outside enemy. The effect of these upon cruisers, even of moderate size, and still more upon a fleet of battleships, is not only the actual injury possible from them, but the delay imposed. This delay indeed is in the strictest sense a strategic factor. As illustrated in the war between Japan and Russia, the outside fleet is enabled to choose its situation within pretty wide limits, in reliance upon the inevitable period of time the inside will need to assure its passage, by determining a safe channel. Yet, while this is true, it is not so unqualifiedly.

Skill and vigilance may now, as in all ages and conditions, enable the one party or the other to get the better; especially the one inside. I presume that a simple application of the three-point problem, to determining a straight channel through a presumed mine field, might be carried on by three lights placed at night for the observers; that such channel might be swept by night as well as by day; and that, once cleared, further laying of mines might be prevented by adequate scouting. Range lights will give pilotage for the channel cleared. Yet, granting that such means may be efficacious, the need for using them and the onerousness of their demands show how conditions have altered in twenty years. Obviously, too, the outsiders must try to stop such operations, with the result of a good deal of fighting corresponding to that which the army calls "outpost."

The ransoms levied by cruisers in 1888, or the alternative bombardment, illustrate both the need of sea-front fortifications for commercial ports, and the needlessness of works on their land side. The cruisers could not have stood up against a very few heavy guns, and they had not force to attempt a landing. For a fleet, or for a great landing army, the game at a mere commercial port would be too small; not worth the candle, as the French say. Such expeditions would direct effort against a naval base. Now that bombardment of unfortified seaports is forbidden by international agreement, the question remains to what extent it will suit the policy of a nation by non-fortification to permit the tranquil occupation of its convenient harbors by an enemy's vessels; for the purpose, for instance, of coaling, or repairing, or demanding supplies. Of course any molestation of vessels so engaged is active war, and would at once deprive the port of the immunity attendant upon not being fortified.

A word is due on the subject of coast-defense ships, although we hear less of these now than once. A floating defense which is confined to the defensive, by which is meant that it can put forth its offensive power only when the enemy sees fit to attack, is inferior to the same amount of offensive power established ashore, for several reasons: (1) Because it cannot bear the same amount of weight as a land work; (2) because it is open to modes of attack to which the land work is not open, as the torpedo and the ram; (3) because the very factor which constitutes its chief advantage, its mobility, is also a source of weakness by necessitating the attention of a large part of the personnel or garrison to the mere handling of the fortification. To this is to be added a consideration, important to my mind, though I have not seen it noticed elsewhere, that a system of coast defense relying mainly upon ships is liable to be drawn in mass to a point other than the enemy's real objective, and so to leave the latter uncovered. Land works are not open to that mistake. Nelson in his scheme for the defense of the Thames particularly and of the southeast coast of England generally, in 1801, wrote a paper which illustrates his comprehensive military genius as really as his more conspicuous achievements do. In this he laid peculiar stress upon the order that the coast-defense vessels, the blockships, as they were called, should on no account be moved under apparent imminent necessity. Their stations had been carefully and deliberately chosen, in quiet consideration; they must not be changed under the influence of hasty apprehension. Permanent works, established in quiet moments on sound principles, have the advantage that they cannot be shifted under the influence of panic. The distributions of the American fleet during the Spanish war furnish interesting matter for study as to the effects of popular fears on military dispositions; of which, for that matter, general history affords many examples.

A moment's thought will show that one mode of coast defense by the navy to which attention is very largely directed nowadays, that by torpedo-vessels and submarine boats, is not strictly defensive in its action, but offensive. For harbor defense, torpedo-vessels are confined almost wholly to an offensive rôle,—the offensive-defensive,—because an attack by a fleet upon a port will usually be by daylight, while torpedo-vessels, in the general scheme of harbor defense, must limit their efforts mainly to the night. The chief rôle of the torpedo-vessels is in *attack* upon a hostile fleet which is trying to maintain its ground near the port.

The great extension and development given to torpedo-vessels since these lectures were written do not seem, as far as experience goes, to have affected the general principles here enunciated; nor in actual war has anything occurred to contradict the conclusions indicated to students of naval matters twenty years ago. Torpedo-vessels, when relying upon themselves alone, have always attacked by night. By day they have merely completed destruction already substantially achieved by the battleships; and this probably is the function that will fall to them in the unusual case of a fleet seriously attacking fortifications. They then may poniard the wounded, especially if left behind by their friends. The increase of size in torpedo-vessels, above the torpedo-boat of first and second classes, in which they began, has brought with it gun armament, as was then predicted; and gun fights between the torpedo-vessels of the opposing sides, much resembling the skirmishes incident to land sieges, were frequent in the operations around Port Arthur. Had the defensive rôle of the Russians fallen to the Japanese, we doubtless should have had more torpedo attack—the offensive-defensive—directed against the outside fleet. Their audacious attempts to block the harbor, by sinking vessels in the channel ways, give assurance that, in the reverse case, similar energy would have been directed against ships attempting to hold their ground near the port.

Defenses, whether natural or artificial, covering strategic points such as coast fortresses, play a very important part in all warfare, because they interpose such passive resistance to the assailant as to enable smaller force to hold in check a larger. Their passive strength thus becomes equivalent to a certain number of men and allows the holder to let loose just so many to join the active army in the field. The defenses of Port Arthur permitted the tenure of the place by a much smaller number of men than the besiegers were compelled to employ in the siege. This evidently signifies that in the field campaign the Russian army was by so much more numerous, and the Japanese by so much fewer. Places so held serve many purposes and, in some proportion, are absolutely necessary to the control of any theater of war. They are as essential to sea as to land war; but, looked upon as conducive to the attainment of the objects of war, they are to be considered inferior to the army in the field. To take an extreme case, a *reductio ad absurdum,* if the number of such posts be so great that their garrisons swallow up the whole army of the state, it is evident that either some of them must be abandoned or the enemy's army be left unopposed. Thus Jomini says, "When a state finds itself reduced to throw the greater part of its force into its strong places, it is near touching its ruin." This received illustration in the war between Japan and Russia. Russia was reduced to shutting up her fleet in Port Arthur and Vladivostok; and persistence in this course, whether by choice or by necessity, prognosticated the ruin which overtook the naval predominance which at the beginning of the war she actually possessed over Japan.

In the sphere of maritime war, the navy represents the army in the field; and the fortified strategic harbors, upon which it falls back as ports of refuge after battle or defeat, for repairs or for supplies, correspond precisely to strongholds, like Metz, Strasburg, Ulm, upon which, systematically occupied with reference to the strategic character of the theater of war, military writers agree the defense of a country must be founded. The foundation, however, must not be taken for the superstructure for which it exists. In war, the defensive exists mainly that the offensive may act more freely. In sea warfare, the offensive is assigned to the navy; and if the latter assumes to itself the defensive, it simply locks up a part of its trained men in garrisons, which could be filled as well by forces that have not their peculiar skill. To this main proposition I must add a corollary, that if the defense of ports, many in number, be attributed to the navy, experience shows that the navy will be subdivided among them to an extent that will paralyze its efficiency. I was amused, but at the same time instructed as to popular understanding of war, by the consternation aroused in Great Britain by one summer's maneuvers, already alluded to, and the remedy proposed in some papers. It appeared that several seaports were open to bombardment and consequent exaction of subsidies by a small squadron, and it was gravely urged that the navy should be large enough to spare a small detachment to each port. Of what use is a navy, if it is to be thus whittled away? But a popular outcry will drown the voice of military experience.

The effects of popular apprehension upon military dispositions were singularly shown during our war with Spain. Popular apprehension, voiced, it was understood, by members of the national legislature, was the cause of dispositions of the fleet which impoverished the needed blockade of the enemy's ports, and which, in face of a more capable foe, would have enabled the Spanish squadron to gain Cienfuegos, where it would have had the support of the main Spanish army. This, with our very small regular army, and the sickly season beginning, would have been a very different proposition from that presented by the isolated Santiago.

This line of thought requires development. Panic, unreasonable apprehension, when war begins, will be found in the same persons who in peace resist reasonable preparation. Unless my information at the time was incorrect, a senator of the United States, who has earned much approval in some quarters by persistent opposition to naval development, was among the most clamorous for the assignment of naval force to the local defense of his own State, which was in no possible danger. In both cases the effect is the result of unreason. "It is better," said a British admiral of long ago, "to be frightened now, while we have time to prepare, than next summer, when the French fleet enters the Channel." The phrase is much more worthy of perpetuation than his other often-quoted "fleet in being."

Where a navy is relied upon for a pure defensive, the demand will naturally follow for many small vessels,—a gunboat policy,—for the simple reason that tonnage put into large vessels cannot be subdivided. Our early single-turreted monitors, being small and relatively cheap, could be numerous. They therefore lent themselves readily to the scheme of a pure defense, widely distributed; the naval analogue of the now discredited *cordon* policy, in which the protection of a land frontier was attempted by distributing the available force among numerous vulnerable points instead of concentrating it in a central position. Any belief that still exists in those monitors, as suited to a general naval policy, will be found associated with the idea of subdivision, one or two vessels to every port. I read, now many years ago, precisely such a project, elaborated for the defense of our Atlantic coast; one, two, or three, single-turreted monitors assigned to each, according to its assumed importance; and this by a trained naval officer. Happily, the last twenty years has seen the conception of a navy "for defense only" yield to sounder military understanding of the purposes of a navy; and that understanding, of the navy's proper office in offensive action, results as certainly in battleships as the defensive idea does in small vessels.

Every proposal to use a navy as an instrument of pure passive defense is found faulty upon particular examination; and these various results all proceed from the one fundamental fact that the distinguishing feature of naval force is mobility, while that of a passive defense is immobility. The only exception known to me is where permanent—that is, immobile—works cannot be constructed to command the surroundings, because of the extent and depth of the water area to be defended. In illustration, I would cite the suggested artificial island with fortification, proposed for the entrance of the Chesapeake. It is contemplated because the capes are too far apart fully to command the entrance. Conditions being as they are, I conceive that to employ coast-defense ships instead of the artificial island would be a mistake; while possibly, if the water were forty fathoms deep, recourse to a floating defense, elaborately protected against under-water work attack, might be unavoidable, because there would be no alternative measure possible.

Such an exception emphasizes the rule. The strictly defensive strength of a seaport depends therefore upon permanent works, the provision of which is not the business of naval officers. The navy is interested in them because, when effective, they release it from any care about the port; from defensive action to the offensive, which is its proper sphere.

There is another sense in which a navy is regarded as defensive; namely, that the existence of an adequate navy protects from invasion by commanding the sea. That is measurably and in very large degree true, and is a strategic function of great importance; but this is a wholly different question from that of the defensive strength of seaports, of strategic points, with which we are now dealing. It therefore will be postponed, with a simple warning against the opinion that because the navy thus defends there is no need for local protection of the strategic ports; no need, that is, for fortifications. This view affirms that a military force can always, under all circumstances, dispense with secure bases of operations; in other words, that it can never be evaded, nor know momentary mishap.

I have now put before you reasons for rejecting the opinion that the navy is the proper instrument, generally speaking, for coast defense in the narrow sense of the expression, which limits it to the defense of ports. The reasons given may be summed up, and reduced to four principles, as follows:

1. That for the same amount of offensive power, floating batteries, or vessels of very little mobility, are less strong defensively against naval attack than land works are.

2. That by employing able-bodied seafaring men to defend harbors you lock up offensive strength in an inferior, that is, in a defensive, effort.

3. That it is injurious to the *morale* and skill of seamen to keep them thus on the defensive and off the sea. This has received abundant historical proof in the past.

4. That in giving up the offensive the navy gives up its proper sphere, which is also the most effective.

II. Offensive strength

The offensive strength of a seaport, considered independently of its strategic situation and of its natural and acquired resources, consists in its capacity:

1. To assemble and hold a large military force, of both ships of war and transports.

2. To launch such force safely and easily into the deep.

3. To follow it with a continued support until the campaign is ended. In such support are always to be reckoned facilities for docking, as the most important of all supports.

It may be urged justly that this continued support depends as much upon the strategic situation of the port and upon its resources as upon its strength. To this, however, must be replied that it was never meant that the division between the different elements which together make up the total value of a seaport was clear-cut and absolute. The division into heads is simply a convenient way by which the subject can be arranged and grasped more clearly. Some necessary conditions will affect, more or less, all three, strength, position, and resources, and will unavoidably reappear under different heads.

1. Assembling. It will be seen that depth of entrance, and the area of anchoring ground for large vessels, are elements of offensive strength. Without depth the largest ships of war could not get in and out, and without great extent the requisite fleet could not be assembled. Depth of water, however, may be a source of weakness defensively, because allowing the entrance of the enemy's heavy vessels. In a port of secondary importance, fitted only to be a base for commerce destroying (*e.g.* Wilmington, N. C.), there would be no gain of offensive strength, but rather loss of defensive, by great depth at entrance.

Suitable ground on shore for the establishment of docks and for storehouses, for the maintenance, repair, and supply of ships, is a necessary condition of offensive strength. That this ground should be so situated as not to be open to injury by the enemy is a condition of defensive strength, and the same is to be said of the anchorage ground. Healthy ground for the encampment or lodging of troops, etc., may be properly included in the elements of military strength, both offensive and defensive. A special instance of opportunity to constitute this feature of camping ground for an expedition was afforded when the site for the navy yard in Puget Sound was acquired. The original commission recommended the acquisition of an extensive area for that reason, among others; but the recommendation was not adopted.

2. *Launching.* To launch a force safely and easily into the deep implies that when ready to start it can go out at once and take up its order of battle in the presence of an enemy, unmolested; favored in doing so either by the hydrographic conditions outside allowing the necessary maneuvering without interference, or by the protection of the port covering the fleet with its defensive power. It is, of course, perfectly conceivable and possible that a fleet may by its own power insure its own freedom of maneuver; but the time occupied in changing from one order to another is always critical, and such maneuvers should be performed out of the reach of the enemy. In order to complete the offensive strength of the place, it should be able with its own means to cover the fleet during such change of formation; beyond that, the offensive strength of the port for this purpose cannot be expected to reach.

This case is analogous to that of an army passing through a defile,—room must be secured beyond to deploy. If the entrance be narrow, the fleet must get outside before being able to maneuver. In this case, the conditions of offensive and defensive strength again clash, for a narrow and tortuous entrance is most easy to defend. It may be interesting to recall that at the time these lectures were written, 1887, a very large, perhaps even a predominant, naval opinion held that the ram would play the most prominent part in naval warfare. From this followed that fleets would approach one another bows on, and that deployment would mean the formation of a line abreast, bringing all bows towards the enemy. It was such a deployment that was primarily in my mind when the paragraph just read was written. Experience and progress have restored to the gun its supremacy; and as a ship is many times longer than it is wide, a greater amount of gun power can be developed along the side than across the beam. The small, single-turret, monitor is the only exception to this. It follows that deployment, from column heading toward the enemy, means now a change of course, by which the broadsides of all the vessels are brought on to the same line, with all the guns training towards the enemy.

If a fleet is able to steam out from port in line abreast, a change of course all together, when nearing the enemy, effects such deployment; but the channels by which harbors are left are usually too narrow for this. Ordinarily vessels must go out in column, and form line by a graduated movement. An outside enemy awaiting such issue would seek to deploy across the exit of the channel, out of range of the forts but within range of the exit, enabled thus to concentrate fire upon the leaders of the column before the vessels following can give support by deploying their batteries.

Belts of submarine mines, laid by the one party or the other, as was largely done by both the Japanese and Russians, may affect the conditions constituted by nature. Submarine mines may be said to introduce artificial hydrographic conditions. The inside party would aim to keep the enemy, by fear of mines, so far distant as to be out of range of its point of deployment; and the effect may be intensified by an energetic use of torpedo vessels and submarines. At Port Arthur, the Russian mines and the apprehension of torpedo attack did fix the Japanese fleet to the Elliott Islands, so that the Russians when they came out had no trouble about deploying.

An outside fleet, on the other hand, would wish by a like use of mines to prevent the issuing fleet from deploying until it had passed beyond support by the shore guns. The Japanese did not attempt this; that is, one-third of their battleship force having been lost early in the war, the exigencies of their case led them to seek the safety of a boom-protected anchorage, rather than expose their armored ships to torpedo attack by remaining continuously close to the port, in order to obtain the advantage of concentration on the head of the enemy's column. Their mine fields, by making the exit dangerous, enforced delay upon the enemy's fleet, enabling themselves to come up before it could escape; but this strategic advantage was not accompanied with the tactical advantage of concentration upon the leading enemies during the critical moment preceding deployment.

All these dispositions—boom anchorage, mine fields, concentration on enemy's leaders—are tactical. My subject is strategy. The excuse for the apparent digression is that the strength of a naval base of operations is a strategic consideration, affecting all the theater of war. Tactical facilities and disabilities are elements of strength or weakness, and as such a general consideration of them falls under the lawful scope of strategy. Mine fields, as used in the latest war, have introduced a new condition, affecting that element of offensive strength in a naval base which has been defined as the ability to launch a maritime force easily and safely into the deep.

These tactical considerations have a further very important bearing upon a strategic question of the gravest order; namely, the proper position for an outside fleet charged with the duty of checking the movements of a more or less equal enemy within a port. Hawke and St. Vincent in their day answered: Close to the port itself. Nelson, more inclined to take risks, said: Far enough off to give them a chance to come out; to tempt them to do so, for we want a battle. The difference was one of detail, for both aimed at interception, though by different methods. It may be mentioned in passing that Nelson paid for the deliberate looseness of his lines by some periods of agonizing suspense, touch with the enemy being lost.

The strategic reply to this question is as sure as historical experience can make it. The intercepting fleet must keep so near the harbor's mouth that the enemy cannot get a start. A start may be retrieved, as Nelson did; but again it may not, and the risk should not be taken. Wireless can do much to retrieve; but wireless needs a sight of the enemy to give it its message. The Japanese solution at Port Arthur—that is, the stationing their main fleet at the Elliott Islands—answered, because the Russians displayed neither energy nor ingenuity. A channel cleared by night sweeping, adequate range lights, a little "D—n the torpedoes," and a dash by night might have transferred the Russian fleet to Vladivostok, granting the Japanese at the islands. The difficulties were no greater than have been overcome often before, and the strategic situation would have been greatly modified. The Japanese recognized clearly that it would have been to them a distinct check. The tactical problem of getting the Russian fleet out of Port Arthur, under the two suppositions of the enemy at the Elliott Islands and before the port itself, would afford a very interesting tactical study, with profitable strategic conclusions.

If a port have two outlets at a great distance from each other, the offensive power will be increased thereby, the enemy being unable to be before both in adequate force. New York is a conspicuous instance of such advantage. If the two outlets, by the Sound and by the ocean, are suitably fortified, an enemy cannot be near both without dividing his fleet into two bodies out of mutual supporting distance. A united hostile fleet cannot command both channels until right before the city, where the channels meet. The same advantage, to a much less degree, is found at Port Orchard, Puget Sound, and had weight with the commission which chose this point for the navy yard. The port of Brest has the same, which with sailing fleets gave a distinct advantage. Wireless telegraphy of course facilitates the movement of the enemy to one entrance from the other, or from a central position; but the gain over former conditions is less than one would imagine. Nelson, fifty miles from Cadiz, learned of the enemy's sailing in two and a half hours by a chain of signal vessels. His chance of intercepting the enemy was as good, perhaps even better, than that of a steam fleet similarly situated, dependent upon wireless. The speed of the escaping fleet under steam would fully counterbalance, probably more than counterbalance, the gain of the outsider by speedier information. Over twenty-four hours were required for the allied fleets to leave Cadiz before Trafalgar.

In order that two outlets should confer fully the offensive advantage claimed, it is necessary that they should be so far apart that the enemy cannot concentrate before one, between the time that the fleet within indicates its intention of coming out and the time when it has formed its order of battle outside. With steam, few ports are so favorably situated; the dependence of sailing-ships upon the direction and force of the wind introduced a tactical and strategic element which can now be disregarded. "Keep all fast," once wrote Lord St. Vincent, "for we know that with a wind to the southward of southeast by south no ship of the line can leave Brest." The analogy of this to the delay in coming out caused by an enemy's mine field is easily seen.

The third element in the offensive strength of a strategic port has been stated as the capacity, after having covered the exit of a maritime force, to follow it with continued support throughout the intended operations.

Obviously, in any particular port, this capacity to support active operations will depend upon the scene and character of the operations. In the war between Japan and Russia, the Japanese dockyards were the scene both of the equipment and refreshment, restoration and repair, of the ships. They thus followed them continuously; stood at their back. The Russian home ports despatched the vessels, but had nothing to do in sustaining them on the theater of war. A fleet equipped at San

Francisco for operations in the Far East would require support nearer than that harbor. Portsmouth and Plymouth were the great Channel yards of Great Britain a century ago, as they still are; but the two ablest English commanders, Hawke and St. Vincent, would not allow their vessels to seek either for support. For supplies, for refreshment of the crews by rest, for cleaning bottoms, for overhaul of motive power, they were sent to Torbay.

All this is simply to reaffirm that for seaports position—situation—is the first in importance of the elements of strategic value. This illustrates again Napoleon's saying, "War is a business of positions." In the War of Secession, the United States ships were equipped in the northern yards, but were sustained in the campaign by nearer bases,—Port Royal, Key West, Pensacola. This is a frequent condition; indeed, with the wide scope of naval operations, the more usual. But it is better that the two processes, original equipment and continuous support, be combined behind the same defenses, where possible. Having in view the increasing importance of the Gulf and Caribbean, increasing because of the increasing imminence of Pacific questions and the near completion of the Canal, it will be pertinent to inquire closely as to whether the northern navy yards adequately meet possible emergencies of the character now under discussion.

To follow a fleet with support means principally two things: (1) To maintain a stream of supplies out, and (2) to afford swift restoration to vessels sent back for that purpose.

"Supplies" is a comprehensive word. It embraces a large number of articles which are continuously being expended, and which must be renewed by means of storeships periodically despatched. It applies also to maintaining the condition of a fleet by a system of reliefs. This involves a reserve, so that ships long out and worn are replaced by fresh vessels, and, yet more important, by refreshed crews. Of the capacity thus to refresh and thus to replace, numerous dry docks are the most important single constituent, because the most vital and the longest to prepare.

Historical instances, by their concrete force, are worth reams of dissertation. The capacity of the Japanese dockyards may have been ample or may not, but by the Japanese government it was felt to be of critical importance that all their armored ships should be docked and restored in the briefest time possible after the fall of Port Arthur. To achieve this was a matter of great anxiety; to be measured by Togo's signal, "The safety of the Empire depends upon to-day's results." One of the factors in the results of that day was that the dockyards which sent out the fleet were able to dock all the vessels in the respite of time gained by Russia's dilatoriness. Thus the Japanese fleet was an assemblage of veterans restored by repose and repair. Clearly it would have been a better condition if the yards throughout the war could have contained as a constant rule two armored ships, docking and resting crews; a reserve to relieve others, and also at critical moments to augment the total strength of the fleet. It is not ideal management to have to clean and repair an entire navy at the same moment; but it was forced upon Japan by the fewness of her armored ships, which required their constant employment at the front. Such a reserve of ships corresponds to the margin of safety of the engineer.

It may occur to some that this capacity to sustain a fleet in its operations falls more exactly under the head of Resources, the last of the three heads under which the elements which affect strategic value of seaports have been summarized. It is true that this capacity is one of resources, yet it may be claimed that its worth is more evident when considered as an element of offensive strength. If the capacity of dockyards be classed under resources, less attention is attracted to the fact that upon that capacity may depend the offensive energy of a war.

Subdivision of a single subject cannot be into compartments separate from one another. Subdivision is not an end in itself. It is a means to exact thinking, and to thinking out a matter more thoroughly, because more systematically than would otherwise be done. Also, a comprehensive summary under heads tends to insure that in a particular decision or choice of a position no consideration will be overlooked.

3. RESOURCES

The wants of a navy are so many and so varied that it would be time lost to name them separately. The resources which meet them may be usefully divided under two heads, natural and artificial. The latter, again, may be conveniently and accurately subdivided into resources developed by man in his peaceful occupation and use of a country, and those which are immediately and solely created for the maintenance of war.

Other things being equal, the most favorable condition is that where great natural resources, joined to a good position for trade, have drawn men to settle and develop the neighboring country. Where the existing resources are purely artificial and for war, the value of the port, in so far, is inferior to that of one where the ordinary occupations of the people supply the necessary resources. To use the phraseology of our subject, a seaport that has good strategic situation and great military strength, but to which all resources must be brought from a distance, is much inferior to a similar port having a rich and developed friendly region behind it. Gibraltar and ports on small islands, like Santa Lucia and Martinique, labor under this disadvantage, as compared with ports of England, France, the United States; or even of a big island like Cuba, if the latter be developed by an industrial and commercial people. The mutual dependence of commerce and the navy is nowhere more clearly seen than in the naval resources of a nation, the greatness of which depends upon peaceful trade and shipping. Compared with a merely military navy, it is the difference between a natural and a forced growth.

Among resources, dry docks occupy the place first in importance: (1) because to provide them requires the longest time; (2) because they facilitate various kinds of repairs; (3) because by the capacity to clean and repair several vessels at once, and so restore them with the least delay to the fleet, they maintain offensive energy.

Dry docks represent in condensed form the three requirements of a strategic seaport. In position they should be as near the scene of war as possible. Strength is represented by numbers; the more numerous the docks, the greater the offensive strength of the port. For resources, the illustration is obvious; docks are an immense resource. In contemplating the selection of a navy-yard site, it is evident that facility for excavating docks is a natural resource, while the subsequent construction is artificial. Evidently, also, a commercial port will supplement these resources in an emergency by the docks it may maintain for commerce, thus exemplifying what has been said as to the wide basis offered by resources developed by man in his peaceful occupation of a country.

CHAPTER V

STRATEGIC LINES

THE STRATEGIC POINTS on a given theater of war are not to be looked upon merely separately and as disconnected. After determining their individual values by the test of position, military strength, and resources, it will remain to consider their mutual relations of bearing, distance, and the best routes from one to the other.

The lines joining strategic points are called by military writers strategic lines. On land there may be several lines, practicable roads, connecting the same two points; any one of which may at different times have different names, indicating the special use then being made of it, as, line of operation, line of retreat, line of communications, etc. At sea, other things being equal, the line that is shortest, measured by the time required to pass over it, is ordinarily the one to be chosen by a fleet; but this obvious remark, approaching a truism, is open to frequent modification by particular circumstances.

Illustration is afforded by the very recent case of Rozhestvensky's fleet when leaving French Cochin China, or, yet more critically, the Saddle Islands, for its final push towards Vladivostok. In the first instance, there was the question of passing between Formosa and the mainland, the direct route, or going outside the island. The latter was followed. In leaving the

Chapter VIII, *Naval Strategy* (Boston, 1911), 164–99.

Saddles, the shortest route, by the Sea of Japan, was the one chosen; yet with all the risks that would be involved in the greater delay occasioned by going east of Japan, through Tsugäru Straits, it is evident that there were favoring chances, which needed careful weighing. The position occupied by Admiral Togo was judiciously chosen to facilitate intercepting in either case; but the fact that he passed through some thirty-six hours of anxious suspense, because the Russians did not appear nor tidings of them come in, shows the possibilities of the situation. The very strength of his conviction that they *must* come that way would be an element favoring the Russians, had Rozhestvensky decided for the other. There are temperaments which cannot readily abandon a conviction, as there are others which cannot bear suspense.

Of the numerous lines which may be traced on the surface of the globe joining two seaports, two general divisions may be made,—those that cross the open sea, and those that follow the coast-line. A glance at the chart of the Gulf of Mexico will illustrate my meaning, showing the two available routes to Key West from the mouth of the Mississippi, or from Pensacola. To use the open sea, which is generally the shortest, military command of the sea is needed; when this command is not held, vessels are forced to follow their coast-line, usually by night, and to use such harbors of refuge or other support as the coast with its conformation will give. The flotilla with which the first Napoleon intended to invade England illustrated this method. The large number of vessels composing it necessitated building them in many different places. To reach the point of assembly, Boulogne, they had to run the gantlet of the British cruisers that controlled the English Channel. This was successfully done, though with a certain proportion of loss, by keeping closer to the coast than the enemy could safely follow, while an elaborate scheme of coastwise defense by stationary and flying batteries was also provided for them. In the War of 1812, American coasting trade, as far as it survived, was driven to the same evasion, but without the same support. The same conditions prevailed in Nelson's time along the Riviera of Nice and Genoa. Whenever the open sea is controlled by an enemy recourse will be had to this means, usually by night; for, while land communication is ampler and surer than formerly, it is not yet able to replace the coasting trade. It is only necessary to consider the coal traffic by sea from the Delaware to New York and the Eastern States, to see that it cannot lightly be surrendered, or replaced by railroad, without much suffering to the community and derangement of industries.

Neutral coasts may thus in some degree be made part of the line of approach to belligerent ports, endangered by the nearness of an enemy. For example, if war existed between Germany and Great Britain, with the British navy controlling the North Sea, German vessels having once reached the coast of France or of Norway might proceed with safety within the conventional three-mile limit.

The most important of strategic lines are those which concern the communications. Communications dominate war. This has peculiar force on shore, because an army is immediately dependent upon supplies frequently renewed. It can endure a brief interruption much less readily than a fleet can, because ships carry the substance of communications largely in their own bottoms. So long as the fleet is able to face the enemy at sea, communications mean essentially, not geographical lines, like the roads an army has to follow, but those necessaries, supplies of which the ships cannot carry in their own hulls beyond a limited amount. These are, first, fuel; second, ammunition; last of all, food. These necessaries, owing to the facility of water transportation as compared with land, can accompany the movements of a fleet in a way impossible to the train of an army. An army train follows rather than accompanies, by roads which may be difficult and must be narrow; whereas maritime roads are easy, and illimitably wide.

Nevertheless, all military organizations, land or sea, are ultimately dependent upon open communications with the basis of the national power; and the line of communications is doubly of value, because it usually represents also the line of retreat. Retreat is the extreme expression of dependence upon the home base. In the matter of communications, free supplies and open retreat are two essentials to the *safety* of an army or of a fleet. Napoleon at Marengo in 1800, and again at Ulm in 1805, succeeded in placing himself upon the Austrian line of communication and of retreat, in force sufficient to prevent supplies coming forward from the base, or the army moving backward to the base. At Marengo there was a battle, at Ulm none; but at each the results depended upon the same condition,—the line of communication controlled by the enemy. In the War of Secession the forts of the Mississippi were conquered as soon as Farragut's fleet, by passing above, held their line of communications. Mantua in 1796 was similarly conquered as soon as Napoleon had placed himself upon the line of retreat of its garrison. It held out for six months, very properly; but the rest of the campaign was simply an effort of the outside Austrians to drive the French off the line, and thus to reinforce the garrison or to enable it to retreat.

Rozhestvensky's movement towards Vladivostok was essentially a retreat upon his home base. The Japanese were upon the line of communication and retreat in force sufficient to defeat him, as Napoleon at Marengo did the Austrians. I think that Cervera was headed into Santiago by the belief or the fear that before Cienfuegos, therefore upon the line of his retreat thither, he would meet a force against which he could not hope for success in the condition of his fleet. That the case was not as he supposed it to be remains a reflection upon the management of the United States navy, the reasons for which reach far behind the naval authorities. The Spanish Minister of Marine stated in the Cortes that Cervera went to Santiago because there was no other place to which he could go. Sampson had been heard of at Porto Rico. The instance illustrates the advantage of two ports on the same frontier,—in this instance the south shore of Cuba,—as well as the effect produced by a hostile force upon the line of communications.

Santiago and Cienfuegos illustrate the advantage of two ports of retreat, as Rozhestvensky experienced the disadvantage of only one. For a fleet acting offensively from a given coast as a base, two ports on that coast also facilitate communications in two principal ways: 1, the raiders of the enemy cannot concentrate on one line, but must divide on two, which halves the danger from them; 2, two ports are less liable to be congested than one is. The question is very much the same as the supply of a division of guns on board ship; how many guns can one chain of supply serve? Napoleon enunciated the following definition: The Art of War consists in dissemination of force in order to subsist, with due regard to concentration in order to fight. To provide two or more ports of supply is to disseminate the means of subsistence without impairing the concentration of the fleet.

Santiago and Cienfuegos—to which may be added Havana, as on a coast line strictly continuous and having land communication with the other two—may be cited as illustrating that a coast line with several suitable ports is essentially one long base of operations, interconnected. By the means at its disposal, torpedo vessels and cruisers, it will be able more or less to keep the immediate neighborhood of the shores free from molestation by an enemy's cruisers. Such a coast-line is therefore a strategic line, embracing several strategic points. It will happen more often than not that several points on the sea frontier nearest the theater of war must be occupied by a state, for strategic reasons. When great efforts have to be made, it will be necessary to carry on the preparations at more than one point. In Napoleon's Egyptian expedition in 1798, France had on the Mediterranean coast only one naval port properly so called,—Toulon; but detachments were prepared in several other ports under French control, and joined subsequently to the sailing of the main fleet. Other reasons may impose a similar distribution of activity. Moreover, it is hazardous to depend exclusively on one arsenal for the supply or repairs of a navy, for a successful blockade or attack might paralyze all operations depending upon it, and the retreat of a beaten fleet upon a single point is more easily intercepted. It is difficult to imagine a more embarrassing position than that of a fleet, after a decisive defeat, hampered with crippled ships, having but a single port to which to return. It may be laid down as an essential principle that on every sea frontier there should be at least two secure ports, sufficiently fortified, and capable of making any and all repairs. In such cases pursuit may be baffled, if the enemy can be dropped out of sight; but with one port he knows to which you are bound. Togo, for instance, knew that Rozhestvensky must be bound to Vladivostok, although he did not know whether he would go through the Strait of Korea or that of Tsugaru. If the two ports are tolerably near each other so much the better, as the enemy cannot then judge the aim of the retreating fleet by slight indications.

Chesapeake Bay and New York on our Atlantic coast are two ports clearly indicated by nature as primary bases of supply, and consequently for arsenals of chief importance. For these reasons, they are also the proper ports of retreat in case of a bad defeat, because of the resources that should be accumulated in them; and both for supplies and as refuges they should be adequately fortified on the land side as well as the sea. Other ports on the same coast, Boston, Philadelphia, Charleston, and others, may serve for momentary utility, disseminating provision and preparation; but the protection given them as commercial ports will suffice for the inferior use

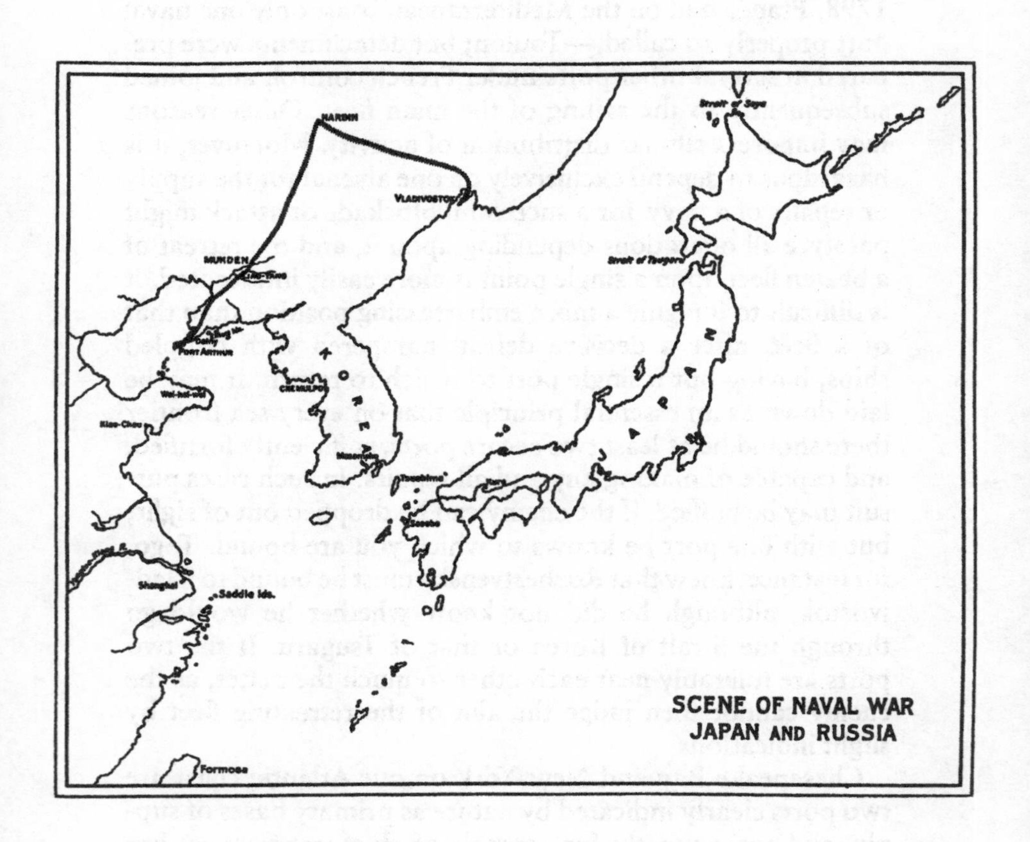

SCENE OF NAVAL WAR
JAPAN AND RUSSIA

made of them for supplying the fleet. Economy of means and of money forbid the multiplication of maritime fortresses beyond the strictly necessary; and it seems probable that the pronouncement of the Archduke Charles, that one first-class and one second-class fortress is sufficient for a land frontier, is true still of maritime fortresses. How it may be for land is beyond my province to say. It can be seen that New York by its natural advantages lends itself profitably to a development greater than Norfolk; because of its two entrances, and because Narragansett Bay could be embraced in the general scheme of defense for New York. This would provide practically three entrances or exits for a fleet.

Vladivostok and Port Arthur illustrate the same propositions, though the situation was for them immensely complicated by the intervention of the Korean peninsula. They were thus related to one another, as to situation, rather as are the ports of our Atlantic and Gulf coasts, with the Florida peninsula intervening; or, even more emphatically, as San Francisco is to Norfolk. In this last case the scale is greater, as it is also in the Russian separation between the Baltic and their Far Eastern ports; but the result is similar. Water communication between the ports is made more difficult by the projection of land, which not only increases the distance between them but affords an obvious strategic position—near the point of Korea, or near the point of Florida, or, in the case of the Atlantic and Pacific, near the Canal—at which a hostile fleet can wait in concentrated force, sure that escapers must come near them. This has been the case in all littoral warfare; capes are the points of danger just as salients are recognized to be. This is the precise opposite of the beneficial effect exerted by Long Island upon New York. The coast of the mainland there is a re-entrant angle, covered by the island as if by an earthwork. This enables the defense to concentrate before going out and embarrasses the enemy by the uncertainties of two exits.

On the same sea frontier all the fortified ports will form

parts of the base of operations, which itself may be properly called a strategic line. Provision should be made for safe and rapid communication between the ports; for while dissemination may be necessary to rapid preparation, concentration is essential to vigorous execution.

In conformity with this statement, of the need to provide for safe and rapid communication between the ports of a maritime frontier, in order to concentrate the forces when the moment for action arrives, we find mentioned among the needs of a base of operations ashore that of free movement and transport of troops and supplies behind the actual front. The river Rhine affords an illustration. In case, as has happened more than once, that the French intended to invade Germany, their army would be on the west side, and they might or might not hold bridges. Whatever the state of the case, they would be able to move their troops, dispersed in cantonments, behind the Rhine, and concentrate them where they preferred, unseen and unknown to the enemy except through spies or treachery. Reconnoissance in force would be difficult, or impracticable. In such movements the river Rhine was called expressively a curtain—it concealed as a curtain does. Long Island Sound will afford similar advantages for the operations of a fleet, if the eastern end be fortified. As has already been suggested, Narragansett Bay could be included in the scheme of a fortified base, the central natural feature of which is Long Island.

As a rule, however, this condition cannot be realized for the ships of a fleet. Their movement from port to port must, ordinarily, be made outside, that is, in front of the base; either directly or by following the coast-line, according to the degree of control possessed over the sea. The internal navigation behind the sea islands of the Southern States does indeed suggest an ideal frontier, in which ships of the heaviest draft could move behind the base from one port to another, as between the two entrances of one port, sheltered from attack by intervening land; but it is only an ideal. Such internal navigation, however, might be used effectually to keep clear a necessary belt of sea outside, by the facilities which it offers for the concentration and sortie of light vessels—of which the torpedo boat or submarine is the most probable—in such numbers as would make the enemy cautious about near approach. The precaution observed by the Japanese to remove their battleships from under-water attack, by holding them at a distance, enforced as it was by two severe disasters, is warrant for believing that torpedo warfare can be so utilized as to assure, in a military sense, the passage of single ships or small divisions from port to port of a threatened coast. The dispositions of the American fleet before Santiago in 1898 are not likely to be repeated.

A like facility for the operation of torpedo vessels would be given by shoals lying along a coast; either by the intricacy of the channels through them, or from their being everywhere impassable for heavy ships. The chain of islands bounding Mississippi Sound, continued by Chandeleur Islands and shoals, establishes a continuous system of navigation for small torpedo boats from the Passes to Mobile. Portions of the Cuban coast present similar features. If the Mississippi and Mobile were two points of the American base of operations on the Gulf, the coast-line joining them could become a fairly secure strategic line of communication—by keeping numerous torpedo vessels moving there—in case the enemy's control of the sea forbade attempting a straight course. East of Mobile this broken ground ceases; but it is probable that the nearness of Mobile to Pensacola would enable the same character of defense to be extended as far as Pensacola.

A coast-line being regarded as a unit, a strategic line, having two or more important strategic points, it is clearly possible that the fleet for various reasons may not always be concentrated. Recent instances are usually most profitable to consider. The fleet may be divided by original mistake, as the

Russians were divided between Port Arthur and Vladivostok. It may be divided by exigencies of preparation, of repair, as for docking, or by accidents of war. A disabled ship must get in where it can. After the engagement of August 10 the Russians so divided. Most returned to Port Arthur; one battleship went to Kiao Chau. After the Battle of the Japan Sea dispersal in several directions took place. The common result of a great victory is such separation of the beaten vessels, just as on land a really great victory is followed by a disintegration, which it is the duty of the victor to increase by the immediate vigor of his pursuit, disorganizing still further the shattered army. The recurrence of such conditions of separation is a permanent feature of warfare, of which strategy and tactics have to take account in all ages. The methods of successive eras will differ with the character of the instruments each has. Sail and steam possess very differing potencies; but the factors in the hands of the opposing parties are, or should be, the same in any particular age. Sails are not opposed to steam, for they do not co-exist. Sails meet sails only, and steam, steam.

The problem of uniting a divided fleet, or of getting a separated ship safely to her main body may therefore be expected to recur; consequently the provision of methods to that end, by using the means of the day, is not of barren academic interest. Nor does the fact that the operation is very difficult, results doubtful, remove the consideration as impracticable. The very improbability of an effort has often been the cause of its success. In the case of a single armored ship, or of a small division, having to run for it in order to effect a junction with the main fleet in another port, the torpedo force could be assembled in such numbers as might be necessary to accompany the passage, which would commonly be made by night; for obscurity is a curtain that favors the weaker. Local familiarity, too, is a much stronger factor than the local knowledge given by charts, especially in the dark. This, and the choice of time,—all the elements, in short,—favor the local navy in such measures. This, however, is not to say that they involve no risk. War cannot be made without running risks. As for the torpedo force employed, all history, including the war between Japan and Russia, affirms the ease with which small vessels can proceed along their own coast, defiant in general of the outsider's efforts, though sometimes caught.

Off-lying obstacles to navigation are of strategic importance and may be looked upon as outworks: generally, however, coming under the head of defensive value. They rather keep off the enemy than facilitate offense. They played a conspicuous part off the coast of Holland in the old English-Dutch wars; but as the size of ships increased, the advantage for defense was more than compensated by the loss of offensive power, the Dutch ships of war remaining smaller and less weatherly than the French and English, owing to the shallower water in which they had to move at home. Consequently, the Dutch line of battle was weaker at any one point than the enemy's force brought against it. The strategic value of the shoals, therefore, introduced an element of tactical weakness in the Dutch navy.

So far, the strategic points of a theater of sea war have been considered only with reference to that particular theater,—to their importance intrinsically, and to their relations to one another and to the fleet. The treatment of the subject would not be complete without a reference to the distance separating colonial possessions or outlying interests from a mother country, and to the effect of that distance upon their value to the holder. This is a branch of the subject which particularly concerns naval war as compared with that on land. The great military nations of the world being found almost wholly on the continent of Europe, with well-established frontiers, the distance of any point defended by them, or against which they move offensively in continental wars, is not very great, at least at first. There is also nothing on the Continent that corresponds to the common ground which all peoples find in the sea, when that forms one of their frontiers. As soon as a nation in arms crosses its land frontier it finds itself in the territory of a neutral or of the enemy. If a neutral, it cannot go on without the neutral's consent; if an enemy, advance must be gradual and measured, unless favored by overwhelming force or great immediate success. If the final objective is very distant, there will be one or more intermediate objectives, which must be taken and held as successive steps to the end in view; and such intermediate objectives will commonly represent just so many obstacles which will be seriously disputed by the defendant.

To push on regardless of such obstacles, and of the threat they hold out against the communications and lines of retreat, requires accurate knowledge of the enemy's condition and sound judgment as to the power of your own army to cover the distance to your distant objective, and to overcome its resistance, before the enemy can bring his own resources into play. This amounts to saying that the enemy is known to be much inferior in strength for the time being, and that you have good hope of striking him to the heart before he is ready to use his limbs and weapons. Thus struck at the very center of his strength, with the sinews of his military organization cut, the key of his internal communications perhaps seized, and concerted action thus hindered, the enemy may, by such a bold and well-timed movement, be brought to submit. This is the aim of modern war, and explains the great importance attached to rapid mobilization.

In naval operations such successes are wrought less by the tenure of a position than by the defeat of the enemy's organized force—his battle fleet. The same result will follow, though less conclusive and less permanent, if the fleet is reduced to inactivity by the immediate presence of a superior force; but decisive defeat, suitably followed up, alone assures a situation. As has been remarked before, the value of any position, sea or land, though very real, depends upon the use made of it; that is, upon the armed forces which hold it, for defense and offense. The sea is not without positions advantageous to hold; but peculiarly to it, above the land, is applicable the assertion that the organized force is the determining feature. The fleet, it may be said, is itself the position. A crushing defeat of the fleet, or its decisive inferiority when the enemy appears, means a dislocation at once of the whole system of colonial or other dependencies, quite irrespective of the position where the defeat occurs. Such a defeat of the British navy by the German in the North Sea would lay open all English colonies to attack, and render both them and the mother country unable to combine effort in mutual support. The fall of any coast position in the Empire would become then a question only of time and of the enemy's exertions, unless the British navy should be restored. Until then, there is no relieving force, no army in the field. Each separate position is left to its own resources, and when they are exhausted must succumb, as did Port Arthur; and as Gibraltar would have done in 1780 but for the navy of Great Britain, which was its army in the field. On the other hand, so long as the British fleet can maintain and assert superiority in the North Sea and around the British Islands, the entire Imperial system stands secure. The key of the whole is held, is within the hulls of the ships.

This is not to say that a powerful, although inferior, navy may not by successful evasion and subsequent surprise seize positions, one or more, in a distant part of the world, and there, so to say, entrench itself, to the discomfiture of the opponent and possibly to the attainment of some distinct ultimate national advantage. The importance yet attached to local bases of operations in remote regions, as for instance by Germany to Kiao Chau in years still recent, might prompt such an attempt. The question then would arise whether the superior naval state would be willing to endure the protracted contest necessary to expel the intruder. General Grant in the spring of 1863 feared that the people of the United States would be discouraged to the point of ceasing the war, if from his operations around Vicksburg he fell back upon Memphis, to take up a new line of advance, which was the course General Sherman urgently advised. This is one of the problems of war, the calculation of chances. Napoleon once said that the art of war consists in getting the most of the chances in your own favor. The superior fleet holds the strongest suit, but the strongest suit does not always win. The character and the skill of the player against you are important factors. For such reasons, the study of the chances, both in general elements of war and in the concrete cases of specific regions, is necessary; in order to fit an officer to consider broadly and to determine rapidly in particular contingencies which may arise.

Readiness and promptitude in action will of course give great advantages in such attempts, as they do in other military operations; and for the matter of that in all affairs of life. There is, however, a recognizable difference between the power of a great state either to attack or to defend a distant and isolated dependency, however strong, from which it is separated by hundreds or thousands of miles of sea, and the power of the same state to support a similar post in its interior or on its own frontier, whether sea or land. The defense of Gibraltar, for instance, would be easier to Great Britain if it were on the British coast. Quebec fell in the Seven Years' War, but during the same period no such mishap occurred to French home fortresses. Rochefort, it is true, might have fallen in 1757 if greater promptitude and enterprise had been shown by the British; but the result would have been of the nature of a successful raid, achieved by surprise, not a permanent tenure consequent upon prolonged operations, as in the case of Quebec.

Other things being equal, the greater the distance the greater the difficulty of defense and of attack; and where there are many such points, the difficulty of defense increases in proportion to their distance, number, and dissemination. The situation of a nation thus encumbered, however unavoidably, is the reverse of that concentration, and maintenance of close communication, which are essential conditions of correct dispositions for war. As was said to Rodney in 1780 by the head of the British Admiralty, the navy cannot be in force everywhere. Some points must be left without the immediate presence of the fleet; and in such circumstances an enemy who has his ships well in hand may by prompt action seize one and so establish himself that he cannot be dislodged. Minorca was thus snatched from Great Britain in 1756, the fleet under Byng failing to dislodge the enemy's fleet before the garrison surrendered; and the French held the island during the remaining seven years of war, although the British navy continued superior. The place, however, could have been recovered by arms, if Great Britain had thought worth while after her fleet had regained full freedom of movement, which it did before the war was over. The end was attained equally surely and much more cheaply by the capture of Belle Isle off the French Atlantic coast, in exchange for which Minorca was returned at the peace. Malta was seized in like manner by Bonaparte in 1798; and, though France had no navy in the Mediterranean, both it and Egypt were held for over two years, when the French were ousted by the British only after prodigious exertions.

The weakness and inconsistency thus brought upon a nation as a whole by the tenure of remote maritime regions or stations must be felt, of course, in due proportion by each of the outlying possessions; which will, by so much, be less secure than equivalent possessions held by a nation whose outposts are nearer to it or less scattered. As compared to the latter, the former is forced to a defensive war at sea, because it has more to lose, the other more to gain; and in accepting the defensive it loses the advantages of the initiative, which are the property of offensive war. This chiefly constitutes the military problem of Imperial Federation, which for several years has been agitating the British Empire. Australia, New Zealand, South Africa, and Canada are self-governing dependencies of the United Kingdom. Each feels that, like Minorca and Malta in their day, it cannot itself alone cope at sea with several possible enemies of Great Britain. Formal independence, as an alternative to the existing self-government, would leave each to its own unaided resources against any one of those enemies; and while it is probable that entire subjugation, that is, conquest and permanent tenure, would be too onerous to attempt, the cession of a particular harbor or district might be exacted, or other commercial or naval advantages, as the price of peace.

Great Britain and France did not besiege Sebastopol because they desired to acquire the place. They attacked there because they thus put Russia at the utmost disadvantage in the matter of communications as compared with themselves, forcing her to defend a maritime fortress at a point distant from the center of national power; as remote then, perhaps, as the recent war in Manchuria has been under the changed circumstances. Having won the victory, they gave back the place, but exacted in return conditions of a different character. The United States did not invade Cuba to acquire the island, but to force Spain to yield conditions not otherwise obtainable. Should the United States have trouble with Japan, and the United States navy be beaten, it is improbable that Japan would seek to annex any part of the American Pacific coast; but she might demand Hawaii, or free immigration of her laborers here, or both. Hong Kong, Kiao Chau, Port Arthur, Formosa, are instances of similar exactions; and the United States' tenure of Guantanamo Bay, though not similarly invidious, illustrates naval strategy availing itself of circumstances in order to obtain advantages of position. The British colonies are thus exposed to be attacked, to the harassment of war, in order to obtain concessions of one kind or another. Under several possible contingencies an enemy's division not only may reach their shores before a British pursuit, but the British may feel it not wholly expedient to pursue, lest the detachment so weaken the home fleet as to render doubtful the security of the British Islands themselves. This is a question of comparative numbers and margin of safety.

Such was the position of Great Britain during the War of American Independence and through the earlier part of the wars of the first French Republic and Empire, and is now. Although other nations, notably France, have very greatly increased the extent and dispersion of their colonial empire as compared with former times, and thereby have multiplied the points at which they are open to attack, their holdings generally have not the economic development, and few of them the commercial value or the national and military importance that attaches to several of the British dominions, colonies, and stations. The very multitude and ubiquity of British maritime possessions, whatever the advantages they have brought with them, hitherto and now, in the way of advancing trade or providing bases for warlike action, were and are a source of danger, of distraction to the defense, and of consequent weakness. There can be no certainty when or where a blow may fall. A French naval officer, speaking of Great Britain's immense naval development alongside of the widespread disposition of her attackable points, has truly said, "England, in the midst of riches, felt all the embarrassment of poverty." The brilliant victories of the Nelsonic period,—the Nile, St. Vincent, Trafalgar,—the overwhelming destruction dealt to the enemies' navies, have obscured the fact that the war, whatever its motive on the part of England, was defensive in its military character, and that to France, despite her maritime weakness, belonged the advantages of the offensive. The British fleets off the French coast stood in the first line of the defense; waiting, longing, it is true, for the opportunity to fight, because in battle they knew was the best chance of destroying the fleets which threatened either their home or their colonies. But still, in attacking, they but defended the country's interests on and across the sea. Their success, however, by the protection afforded to the entire Empire, emphasizes the fact that the supremacy of the fleet was in itself the tenure of the decisive position.

Lord Kitchener, in his visit to Australia and New Zealand in 1910, is quoted as writing in a memorandum to the local governments,

> It is an axiom held by the British Government that the Empire's existence depends primarily upon the maintenance of adequate and efficient naval forces. As long as this condition is fulfilled, and as long as British superiority at sea is assured, then it is an accepted principle that no British Dominion can be successfully and permanently conquered by an organized invasion from over-sea.

But in applying this principle to Australasia, he remarked that considerations of time and space cannot be disregarded. He showed that concentration of force in one or other theater may be compulsory for the navy; that in other seas (than that of the concentration) British naval forces may remain temporarily inferior to those of an enemy, and that some time may elapse before the command of these other seas can be assured. He considered it therefore the duty of all self-governing dominions to provide a military force adequate to deal promptly with an attempt at invasion, and thus to insure local safety and public confidence.[1] The whole argument applies with equal force to a community of self-governing States like the American Union, wherein the Atlantic and Pacific coasts, not to speak of outlying responsibilities, are separated by distances quite as determinative as that between Great Britain and Australasia.

1. *The Mail* (Tri-weekly Times), April 18, 1910.

It will be observed that the successful British strategy of former days consisted in stationing competent divisions of the fleet before the enemy's dockyards. Thus Antwerp, Brest, Rochefort, Toulon, with the intervening Spanish ports when there was war with Spain, indicated a strategic line of operations occupied by the British navy with twofold effect. The occupation prevented the juncture of the enemy's divisions from the several ports; thus stopping concentration, the great factor of effect in war. This result was defensive, and for the whole of the then existing British Empire; the colonies as well as the United Kingdom. Offensively, these main positions covered and supported a blockade of the whole hostile coast. I recall to you these well-understood conditions, in order to draw attention to the fact that, now that Germany has taken the place of France and Spain as the dangerous naval power, exactly the same conditions are found to recur. The British fleet is concentrated in the North Sea. There it defends all British interests, the British Islands, British commerce, and the colonies; and, offensively, commands Germany's commercial sea routes.

If we take a particular very striking instance of sudden seizure of an important position, Bonaparte's Egyptian expedition, which may fitly stand as a type of numerous others directed against England or her colonies, it is seen at once that France is on the offensive, England on the defensive, notwithstanding the brilliant attack and complete victory won at the Nile; the most complete, probably, in the annals of naval war. Bonaparte's phrase to his army, "Soldiers, you are one of the wings of the army of England,"—that is, of the army meant ultimately to invade and reduce England,—was pregnant with truth; nor is there good reason to doubt the reality of his intentions directed against India, or that there were fair military chances of success.

In perfect keeping with his bold system of making war, he had marked the decisive point and pushed directly for it at a moment when there was a strong probability that the enemy would not be ready to stop him before he had reached and seized his object. At the time he sailed, the British had only three ships-of-the-line in the Mediterranean. The reason of this bareness of British force was that France did not stand alone, but had been joined by Spain. The superiority in numbers of the combined navies, localized in the Mediterranean, had compelled the British fleet to withdraw and concentrate in the Atlantic; producing a situation similar for the moment to that indicated by Lord Kitchener, and followed ultimately by the results predicted in his memorandum, when the British fleet, having established control in the Atlantic, returned to the Mediterranean. The condition would be reproduced if Austria now should enter into an offensive alliance with Germany when in war with Great Britain. I mention this because people are prone to think that with steam and wireless and all modern inventions the past cannot recur in essential features; all of us concede that it cannot recur in details. From 1793 to 1795 Spain was in alliance with Great Britain; from 1796 to 1800 she was her enemy. Austria is not now the enemy chiefly feared by Great Britain; but it will be to Austria's interest to see Great Britain out of the Mediterranean, for Austria has great inducements to acquisition within it. Austria and Germany cannot be said to have common objects; but they have a common interest in supporting one another, and their particular objects will be best furthered by coöperating with each other in world politics.

The geographical position of Egypt has given it always unique strategic value, and its political condition in 1798 made a successful seizure in every way probable. Situated at the crossing of many roads,—by land and sea,—opening to Europe by the Mediterranean, and to the Indian Ocean by the Red Sea, a moment's thought will show that Egypt holds to the East and West a position like that which the defile of the Danube held to the battle-ground between Austria and

France, or the Valtelline passes to the Spanish communications through Germany to the Netherlands in the seventeenth century; in a word, that upon political control of Egypt might well depend the control of the East by a nation of western Europe. To strike at India itself was not at once possible; but it was possible to seize, in Egypt, one of those intermediate objectives before alluded to, and there wait until so securely established as to be able to push on further. As the Archduke advancing from Bohemia would secure first the valley of the Danube and then move on to the Rhine, so, advancing against India, France would first seize Egypt and then advance towards the East. Between Egypt and France there was then another important point, another intermediate objective, Malta; to take which Bonaparte paused on his way, notwithstanding the need for haste. The final failure of the expedition must not be allowed to obscure the fact that France was attacking Great Britain, that the latter was doubtful of the object of the expedition and distracted by the numerous points she had to cover, and that French control was successfully established and maintained for a measurable time in the two most important points, Egypt and Malta.

It may be mentioned here that, although in so narrow a sea as the Mediterranean, the greatest perplexity is shown in the correspondence of Lord St. Vincent, the British commander-in-chief, and of Nelson, as well as of the Admiralty, as to the aim of the expedition. Naples or Sicily was thought most probable; and in one of his letters Nelson says, "Malta is in the direct road to Sicily," explaining that it would be most useful as an intermediate base.

So far, this attempt against India, Great Britain's greatest and most distant dependency, had succeeded. Here, however, the difficulty of the French enterprise began. It proved feasible to advance to and accomplish the end in view, if once clear of the harbor and the enemy; but when this had been done no fatal injury had yet been dealt the British, and the French, as towards Great Britain, were forced from the offensive to the defensive. Their conquest must be secured, and its communications with home established, if it was to be effective for further progress. Bonaparte's plan had been sagaciously drawn on the lines of a military operation; it broke down at the point wherein its conditions differed from those of land warfare. Bonaparte, to quote a French author, never attained "le sentiment exact des difficultés maritimes." The army had advanced into the enemy's country; it had seized its first objective; but the blow was not fatal, and its own communications were in deadly danger. There was no relieving force to throw in supplies and reinforcements, as to Gibraltar twenty years before, because the hostile navy controlled the intervening country—the sea. By a combination of genius and good fortune, France had projected its military power to a great distance across the sea and had seized two distant and defenseless stations on a great highway. Could she keep them? We know she did not; the probabilities are she could not; yet she did hold them so long as to justify the attempt made. Once in Egypt and Malta, the French force passed from the offensive to the defensive. The troops in these two outposts became garrisons with no army in the field. The communications between them and with home were closed; and however long the occupation might endure, it was fruitless, except as a diversion to the enemy's forces, unless the enemy wearied of the strife, which at one time already Great Britain had done. Although the unlucky result was hastened and plainly foreshadowed by the Battle of the Nile, it probably was in any case inevitable, in the respective conditions of the two navies, which Bonaparte failed to realize. The whole undertaking from beginning to end illustrates Lord Kitchener's comment on present-day conditions. There is the enforced absence of the British navy due to contemporary military and naval conditions, occasioned by the events of the war in the years immediately preceding, and there is the disastrous ultimate result as soon as the superior navy recovered its freedom of action.

There is also an instructive analogy of outline between Bonaparte's Egyptian expedition and his celebrated land campaign in Italy, 1796, two years earlier. In 1796 he advanced with similar celerity from the Riviera of Genoa, one hundred and twenty miles, to the line of the river Adige, which, with its controlling fortress Verona, he reached and had secured within two months. From several conditions, mainly topographical but not necessary here to enumerate, the Adige with its bridgehead Verona constituted a strategic center, intermediate between the Riviera of Genoa whence Bonaparte started, and Vienna which was his ultimate objective. It thus bore to the campaign the relation which Malta bore to Egypt; but the natural advantages of the position were qualified by the artificial condition constituted by the fortress Mantua, west—that is, in rear—of the Adige, which was occupied by a very large Austrian garrison. While this held out, Bonaparte's tenure of the Adige region was incomplete and insecure. Hence his progress was arrested, as at Malta, by the necessity of mastering Mantua, which flanked his line of advance into Austria, just as Malta flanked the line to Egypt. Mantua detained him eight months, but he maintained a controlling army in the field, as Great Britain at the later period had the controlling navy in the Mediterranean; the mobile force in each case ensuring the communications. When Mantua fell he resumed his march, as he resumed his voyage after Malta; but the success he afterwards achieved, momentous as it was, he himself attributed in part to the fact that Austria weakened, which Great Britain did not do as to Egypt, although some of the British representatives did. "Had the Austrians instead of making overtures for peace continued to retreat," Bonaparte said, "they might have worn down my force." That is just what happened in Egypt and in Malta. The French force in them was worn down.

Since these lectures were written the war between Japan and Russia has afforded a similar instance. Port Arthur was the Malta of Russia and the Mantua of Japan. Russia in the strategy of peace had projected her national power to Port Arthur, far distant from the center of her strength, and there established herself. When war came she was unable to sustain the communications, by either land or sea, as the French in Malta. The place therefore ultimately fell; but the menace of the fleet within to the communications between Japan and Manchuria necessitated the reduction of the fortress, to effect which there had to be a very large detachment from the force available against the main Russian army in the north. From the beginning of the war the Japanese advanced rapidly till they reached Liao Yang, but there they were held for six months, mainly by the siege of Port Arthur, advancing but thirty-five miles. After the fall of the fortress they made preparations to resume their advance, as Bonaparte after Mantua; but the Russians intended to retire, and were collecting their strength as they fell back upon their base. At this conjuncture mediation took place. The difficulty before the Japanese was the same as before Bonaparte when he was advancing towards Vienna; and though they did not sustain the bluff which he did, they acted as he counselled the Directory, "Don't overreach yourselves by grasping at more than the conditions warrant."

The failure of France to maintain her hold in Egypt and Malta, which she had conquered, as well as in other distant points which she held before the war, was paralleled by the inability of Great Britain to keep her colonies in the War of American Independence; not merely the colonies on the Continent but in the West Indies, and to some degree in India and Africa, as well as in Minorca, which fell after a six months' siege. This failure was due to the distance, to the number and distribution of the several positions, which led to scattering the forces, and to the fact that, despite her fine navy, the alliance of the French and Spanish navies was thought, erroneously, to be too strong for her. The extent of the losses of either nation at different times shows the large deductions that must be made from the strength of a strategic position, even as affecting the immediate theater of war, in consequence of its distance from the home country. The great advantage of nearness to the latter is apparent in itself and from these instances. This was the greatest advantage of Japan over Russia in their recent war, and is the advantage which Japan still possesses over all other nations for action in the western Pacific. The center of her national power is close by the scene of possible international contentions in that which we know as the Far East. This will be the advantage possessed by Austria in the Mediterranean should she succeed in pushing her political tenure through the western Balkan peninsula to Salonika and the region round about,—a progress resembling that of Russia through Siberia and Manchuria to Port Arthur, and feared by Italy and Russia, whose jealousy of such a future was shown by the interview of their sovereigns at Racconigi.

Rapid distant expeditions, then, are more feasible by sea than by land, because of the greater mobility of navies; but they are also less decisive in their effects than an equal success won in the mother country or over the fleet, because the blow is delivered upon the extremities and not at the heart. They are also harder to sustain than to make. Once launched and away, if the secret has been well kept there may be good hope of a pursuing fleet taking a wrong direction, though scarcely of the objective being surprised in these days of cables and wireless. But when the immediate object of the expedition has been accomplished, the assailant passes from the offensive to the defensive, with its perplexities; and to maintain his conquest he must control the line of communications, that is, the sea.

It should be noted, also, that for the immediate success

which is essential to final success, such distant maritime expeditions can hope only when there is no effectual opposition to be feared at the point of landing; while ultimate success depends upon there being no interference by the enemy's fleet after landing. This was shown at Sebastopol and at Port Arthur. In neither of these instances was adequate opposition to the landing made, and in neither did the enemy's fleet afterwards trouble effectively the communications of the besiegers. During the War of American Independence, although France and Spain took many small islands from Great Britain, as well as the distant and weak Pensacola, they did not succeed against Jamaica or Gibraltar. At Minorca the result was different; because, although to reduce the island required six months, the British could spare no fleet in that quarter to intercept the expedition, as they did for Jamaica, or to molest the siege, as at Gibraltar. Jamaica was saved by Rodney's victory over the French fleet at the moment it sailed for the attack, in result of which no landing could be attempted. At Gibraltar resistance could not be made to the enemies establishing their lines on land, for they controlled the land; but the British fleet continually interfered by throwing in supplies, and the siege consequently was unsuccessful.

All consideration goes to show that the supreme essential condition to the assertion and maintenance of national power in external maritime regions is the possession of a fleet superior to that of any probable opponent. This simply reaffirms the principle of land warfare, that the armies in the field, not the garrisons, are the effective instruments of decisive war. The occupation of harbors militarily secure, although valuable and even necessary, is secondary to the fleet. Having in view the particular question now interesting us,—the possession of strategic positions in remote regions,—and accepting fully Napoleon's maxim that "War is a business of positions," we may safely coin for ourselves the strategic aphorism, that in naval war the fleet itself is the key position of the whole.

Pertinent to this, it may be noted that the Japanese in the recent war began by landing much of the supplies of the fleet at their protected permanent base in the Elliott islands, but later, as an administrative expedient, found it better to keep a large part afloat. That which is afloat can be kept in vessels capable of accompanying the fleet, which thus carries its base with it, and so can occupy a convenient harbor, though unfortified, its own strength affording for the moment the necessary protection. Efforts for maritime military efficiency therefore must be concentrated on the fleet; but at the same time, as a matter of correct professional thinking, let us avoid the extreme of the Blue Water School, and bear in mind that a fleet charged with the care of its base is a fleet by so far weakened for effective action—weakened both strategically and tactically.

Fortified bases of operations are as needful to a fleet as to an army, but the selection and preparation of them must be governed by certain evident principles.

First, the number of points to be seriously held must be reduced as much as can be, so as to drain as little as possible the strength of the mother country, and to permit her to concentrate on those of vital importance; all others must take their chance with guns pointing seaward only. If the enemy be wise, he will not waste time and strength on them. On the other hand, the vital points should be most seriously strengthened and garrisoned. If the enemy take the offensive he does so against the whole system, and each point that is attacked should be prepared to hold out for the longest time its natural advantages will permit. Every day it does so is gained for the common defense. A very serious effect might have been produced on the Union forces and general campaign if Forts Jackson and St. Philip, below New Orleans, at that very critical period of the war, in 1862, had held out as long as they might have done. The resistance of Port Arthur weakened seriously the Japanese main advance, by the number of men necessarily employed in the siege, and so gained strength and time for the whole Russian scheme of operations. If the early stages of the resistance had been more successful in holding the besiegers at a distance, Rozhestvensky when he arrived might have found the Port Arthur fleet still in existence. The French garrison of Genoa in 1800 marched out an array of skeletons, but their hardihood had gained Bonaparte the time to place the army in the field across the Austrian communications with home. On a smaller scale, Ladysmith played a similar rôle in the Boer War.

A nation that has numerous scattered maritime positions should therefore carefully study how many she can maintain, and which they should be; while, on the other hand, one which sees a necessity arising for establishing her power, or preparing for its future assertion, in a particular region, should as diligently inquire what directions her efforts should take for securing strategic and tenable ports. This, for instance, Germany recently did in the instance of Kiao Chau, and the United States in those of Hawaii and Guantanamo.

Second, there is an evident order of consequence among the various ports which may constitute the maritime system of a particular nation. In the case of all states the home ports come first; because the possibility of a country being thrown on the defensive always exists, and self-protection not only is the first necessity of a nation, but constitutes also the basis upon which alone can rest external action, near or remote. Not till national power is consolidated at home can expansive activity take place. To assuring self-defense succeeds the maintenance of the national policies, in their relative degrees of importance. As these may vary from age to age, the value of ports will vary also. Nevertheless, at any particular epoch there will be a national policy more or less clearly formulated; and the remoter ports, which are essential to the fleet's part in this, should be regarded together with the home ports as a whole, a system, not merely as isolated positions.

Thus, to take the chief maritime state of modern history,—

Great Britain. At the opening of her real career as a naval power Holland is the enemy, and the great dockyard is at Chatham. To this, now that Germany has become the rival naval state, the new position Rosyth, with Chatham, correspond. To antagonism against Holland succeeded alliance, the two states sharing in the universal combination against Louis XIV. Military policy then drew Great Britain to the Mediterranean, whither commercial interest had already drawn her navy in support of the merchant ships. The occupation of Tangier and its development by fortification and mole were an abortive first fruit of British interest in that sea; but the successive acquirements of Gibraltar, Minorca, Malta, emphasized that the Mediterranean had become the first object of Great Britain's policy, after self-protection in home waters was provided. All these three were fortresses in the strict sense of the word.

As the eighteenth century advanced, British interests in the Mediterranean remained, but became secondary to those in the West Indies and in North America. The business of the old "Turkey merchant," as he used to be called, took rank beneath the sugar of the West Indies, the rice and tobacco of the American continent, the furs of Canada, and the fisheries of Newfoundland. Jamaica, which in my judgment has the most controlling situation in the Caribbean, we may infer to have been strongly fortified and garrisoned, from the extent of the preparations for its reduction made in 1782 by the allied nations. France had become the enemy, and so remains through the century. This condition is emphasized at home by the growing importance of Portsmouth and Plymouth as dockyards; but the continental colonies of Great Britain, now embraced in the United States, seem to have had only seacoast defense. This is a tacit recognition that they are already too strong in population, as in extent, and too distant, for conquest by a foreign nation, provided the British navy maintained the superiority at sea which it had throughout. Louisburg and Quebec are fortified by the French and garrisoned against siege, just because the population of Canada is so little numerous, and the French navy so inferior, that neither by land nor by sea is their security assured. Their fall emphasizes one consideration in fortification too easily overlooked; namely, that a fortified place, when it passes into the hands of an enemy, transfers to him the advantage, not only of the situation, but of the strength of the works also. If a colonial port thus falls, it is to be desired that it should not also afford immediate artificial protection against recapture from the land side; as Quebec, for instance, did in 1760, in the winter following Wolfe's victory. The deduction from this is, that in places which justify fortification both the works and the garrison must be adequate to all probable exigencies.

During the periods mentioned, British national policy developed coincidently with almost constant war. Hence, the scheme of fortified stations rather grew than was studied; in this much resembling the British Constitution. The United States has had little war, and her external policies have developed unaffected by that military atmosphere which insures unconscious preparation. As regards conditions changing, it may be interesting to recall that when the lectures which constitute the body of this treatment were first written, the armored fleets of the United States, of Germany, and of Japan, did not exist; that Cuba, Porto Rico, and the Philippines were still Spanish, and Hawaii an independent community. It may be added that the General Board of the navy had not been constituted, nor the Joint Board of Army and Navy Officers.

It would be inappropriate to discuss here in detail a scheme of fortified ports for the United States, for the obvious reason that the boards just mentioned are doubtless dealing with the matter, with a thoroughness and an extent of information not here available. But some general strategic considerations may be summarized.

First of all, what is the essential military requisite of a naval station? Evidently that it should be useful in war. Now, in these days, when it takes at least two years to build and equip a battleship, it is evident at once that shipbuilding cannot be reckoned a primary military object in a navy yard. If ships built in a navy yard are better, or cheaper, or built more rapidly, those are good industrial or economical reasons; but none of them is a military reason. The highest function of a navy yard is to maintain the fleet in efficiency in war; and especially to restore it in the shortest possible time when suffering from injuries, whether arising from ordinary service or in battle. No utility in peace will compensate for the want of this in war. The selection of particular sites to serve this end should be governed by this one consideration, of usefulness in war; which may be analyzed, as we have before, into Position, Strength, and Resources. Of these resources, the chief one is copious provision for docking rapidly; and a site that lends itself to this is to be preferred to one that does not, even if there be some advantage as to situation or natural resources. Evidently the differing degrees in which the three requisites may exist will complicate decision, but it need not and should not be complicated by considerations of building ships.

This applies to all chief naval stations, whether on the home coasts or abroad. At home each coast frontier should possess two such naval stations; one of which may be chief, the other secondary in development. As regards stations external to the home country, the number and choice of them depends upon the national policy. If such policy fasten on interests near home, as in the Caribbean, the development of a naval station there may be conditioned by the proximity of the home ports. In the War of Secession, for example, Port Royal, Key West, Pensacola, and New Orleans were all naval stations, but of very limited development. The character of the war, the enemy having no fleet, allowed vessels to be sent to the Northern dockyards for repair, the force at the front being maintained by reliefs.

Some system of reliefs is needed for every force; but it will be realized that docking at least should be possible at a less distance than from Mobile to Norfolk, or to New York, and this facility must be insured, when attainable. The War of Secession was one of very numerous vessels of moderate size, little homogeneous, and essentially not concentrated except on rare occasions of battle; the time for which was at the deliberate choice of the one side possessing a navy. Where fleet is opposed to fleet, each of a limited number of large ships, it will be very urgent that a vessel or vessels spared for repairs should not have far to go nor long to wait.

When these lectures were first written the United States had but one external policy, properly called policy,—the "Monroe Doctrine." She now has two, the second being the "Open Door." Doubtless external relations bring up many kinds of questions, the treatment of which by the Government proceeds for the most part on certain established principles. These principles, because ascertained and fixed, may be styled policies; but they apply to special and occasional incidents, and therefore have not the continuous influence attaching to the two named. These depend upon conditions so constant and so determinative of national attitude, as well as essential to national well-being, that they have formative effect upon national opinion and steady influence upon diplomacy. The Open Door, which in usage indicates equal commercial opportunity, is intimately associated locally with the question of Asiatic immigration to America. Asiatic immigration again is closely linked with the Monroe Doctrine, for it has become evident that Asiatics are so different from Europeans that they do not blend socially. They live side by side, but as separate communities, instead of being incorporated in the mass of the population. Consequently a large preponderance of Asiatics in a given region is a real annexation, more effective than the political annexations against which the Monroe Doctrine was formulated. Hawaii is an instance in point; and the well-known objections of Japan to the political attachment of Hawaii to the United States would undoubtedly have gone further, if more imminent questions had not commanded her attention. Free Asiatic immigration to the Pacific coast, in its present condition of sparse population, would mean Asiatic occupation—Asia colonized in America. This the United States Government cannot accept, because of the violent resistance of the Pacific States, if for no other reason.

This combination of facts, resulting in a national policy, imposes naval stations in the Pacific, just as the entrance of the Mediterranean into the sphere of British interests compelled the gradual acquisition of stations there. While the Monroe Doctrine was the sole positive external policy of the United States, as contrasted with the negative policy of keeping clear of entanglements with foreign states, national interest gradually but rapidly concentrated about the Caribbean Sea; because through it lie the approaches to the Isthmus of Panama, the place where the Monroe Doctrine focusses. This was the condition when these lectures were first written. The question of the Pacific and its particular international bearings was then barely foreshadowed and drew little attention. Now, for the several reasons stated, the Pacific possesses an actual immediate importance; indeed, the shifting of interest may be compared to that which in the latter half of the eighteenth century made the western Atlantic, from Canada to Venezuela, overpass the Mediterranean in the appreciation of British statesmen. The Mediterranean did not thereby cease to be important; it only lost the lead. In the same way the Caribbean remains important; perhaps it has not even quite lost the lead, but it is balanced by the Pacific. The approaching completion of the Panama Canal will bring the two into such close connection that the selected ports of both obviously can and should form a well-considered system, in which the facilities and endurance of each part shall be proportioned to its relations to the whole.

Finally, the maintenance of any system of maritime fortified stations depends ultimately upon superiority upon the sea—upon the navy. The fall of a wholly isolated strong post may be long postponed, but it is sure to come at last. The most conspicuous instance is the celebrated three years' siege of Gibraltar, from 1779 to 1782. All attacks against the Rock were shattered to pieces; but it must have fallen, save for the energy and skill of the British navy in throwing in supplies. The active army in the field, let us say, relieved the besieged fortress.

An immediate corollary to this last proposition is that in war the proper main objective of the navy is the enemy's navy. As the latter is essential to maintain the connection between scattered strategic points, it follows that a blow at it is the surest blow at them. There is something pitiful in seeing the efforts of a great naval force, with the enemy's fleet within its reach, directed towards unimportant land stations, as was the case with the French fleet under D'Estaing in the West Indies during 1778 and 1779; or even against important stations like Gibraltar, to the exclusion of the hostile fleet. The service of the fleet and of the ports is reciprocal; but, except the home ports, they have more need of it than it of them. Therefore the fleet should strike at the organized force of the enemy afloat, and so break up the communication between his ports.

CHAPTER VI

DISTANT OPERATIONS AND MARITIME EXPEDITIONS

NOTWITHSTANDING the difficulty of maintaining distant and separated dependencies, a nation which wishes to assure a share of control on any theater of maritime importance cannot afford to be without a footing on some of the strategic points to be found there. Such points, suitably chosen for their relative positions, form a base; secondary as regards the home country, primary as regards the immediate theater.

The principle laid down by military writers, that an army advancing far from home should establish a second base near the scene of operations, on the same principles that determine the character of the first, and with sure communications knitting the two together, holds good here; only it must be remembered that secure communications at sea mean naval preponderance, especially if the distance between the home and the advanced bases be great. Such secondary bases should be constituted on the same principles as those of the home frontier; that is, it is expedient that there be two fortified ports, of which one only need be of the first order. They should be near enough to yield each other support, but not so near as to allow the enemy to watch both effectively without dividing his main fleet. In 1803 to 1805 the British fleet under

Chapter IX, *Naval Strategy* (Boston, 1911), 200–42.

Nelson, watching Toulon, thus had at its disposal both Malta and Gibraltar. These not only shared in supporting the fleet, but each supported the other by dividing the burden of protecting the long line of commercial communication from the Channel to the Levant. If the Russians in the years preceding the late war had sent their entire fleet to the Far East they would have outnumbered the Japanese and rested upon Port Arthur and Vladivostok. The Japanese, on the other hand, would have had the advantage of the Inland Sea and its several exits for combining unexpected movements.[1] A port like Kure, on the Inland Sea, with two or more entrances widely removed from one another, has the advantage of two ports combined with the advantage of activities concentrated in one. When two ports are possessed, as in these instances, the base of operations comprising two or more points may be thought of as a line, like the home coast frontier. The ideal condition is that the ports should be in communication by land as well as water. Ports in the large islands, Cuba, Haïti, Jamaica, would possess this advantage, for example, Santiago, Cienfuegos, Havana; but two ports in any one of the Lesser Antilles would be too near together. They would be practically a single port.

If the given theater of maritime war be extensive and contain many points susceptible of strategic usefulness, the choice among them becomes important. If a point be central its influence is more evenly distributed, and from it all parts of the theater are more easily reached; but if its influence does not extend to the boundary lines of the area in question its communications with home are endangered. Thus Jamaica, from its central position, is one of the most important points in the Caribbean; but if Great Britain were confined to Jamaica the communications from home, passing through the passages controlled by other nations, would be insecure. The

1. See map on page 148.

same could have been said in 1798 of Egypt, relatively to France, though central as regards Europe and India, if without Malta or an equivalent; and indeed, at times, notwithstanding the commanding position of Egypt, the British have felt uneasy as to their communications with it, although they have Gibraltar and Malta, both secure against a *coup de main* and giving shelter to their fleet. First in order among external positions are those lying on the side nearest the mother country; only when safe here can a farther step be sure. Gibraltar, for instance, may be considered a necessary first step to Egypt; Santa Lucia, a convenient half-way house, at the least, to Jamaica. Second in order, though first in importance on the particular theater, are the central positions; for example, Malta in the Mediterranean and Jamaica in the Caribbean. Those lying on the side farthest from the nation interested, however important, are most exposed, for example, Egypt and Panama, and care should be taken to strengthen their communications with home by intermediate posts. Great Britain has a chain of such posts to India.

From all these considerations, it follows that when a government recognizes that the national interests in a particular region may become of such character as to demand military action, it should be made the business of some competent body of men to study the ground carefully, after collecting the necessary information, and to decide what points have strategic value and which among them are most advantageous for occupation. When such positions are already occupied, the tenure of the present possessor has generally to be respected, and the conditions under which it becomes right to disregard it are not within the decision of the military man, but of the statesman. It will be granted, however, that occasions may arise in which a state may exercise its rights of war in order to protect interests which it thinks vital; and that the control of a maritime region may become a necessity of the war, if not its prime motive. When this is the case, what are technically called "operations of war" follow. The state may aim either at acquiring control, or at extending the control it already has; or, on the other hand, may seek only to defend that which is in present possession, by checking advances threatening to it.

If the aim be to acquire control not already held, the war becomes offensive in its motive and, necessarily, in its operations also. The military operations, however, may not be directed immediately against the object the acquisition of which is sought. It may be that the enemy is more assailable in some other point which at the same time he values more; and that by moving against this, the true object may be more surely reached than by a direct attack. The question here touches the entire conduct of operations. It is, for the belligerent government, analogous to that which presents itself to the commander-in-chief before every position from which he seeks to dislodge an enemy,—shall it be attacked in front, or turned? The former takes more strength, the latter more time. An illustration is to be found in the attempt of the French and Spanish to take Gibraltar from England during the War of American Independence. The attack was made direct upon Gibraltar—the strongest military post in the British dominions—and failed. The same amount of power directed against the English Channel and coast, and skilfully used, could scarcely, under the existing conditions of immense numerical naval superiority, have failed to wring from England the cession of Gibraltar. The conquests of a war are frequently valuable only as a means of barter in the treaty that ends it. The correspondence of the first Napoleon teems with instructions to this purport.

It may thus happen that the object of the war may not be the objective of the military plan. The object of the war, indeed, may not be the gain of territory at all, but of privileges or rights denied before; or to put an end to wrongs done to the declarant. Even so, an attack upon some of the enemy's possessions will probably form part of the plan of operations.

The case before us is limited by our subject to the control of a maritime region,—a control to be either partial or total according to circumstances. To embark upon such a war with any prospect of success, a nation must have two conditions: first, frontiers reasonably secure from vital injury; and secondly, a navy capable of disputing the control of the sea with its enemy under his present conditions. The frontier, or coast, in its broadest sense, is the base of the whole war, the defensive upon which it rests, answering to the narrower base of operations from which a single operation of war starts. The navy is the chief arm by which the offensive is to be carried on; for, while in the defense the navy plays a secondary rôle, in offensive naval war it takes a leading place. Even in case of a large *combined* operation, the chief part is reserved to the ships; unless under circumstances when the enemy has none, and the work of the fleet is so found done to its hand, as at Sebastopol. To that scene of war there were two lines of communications: one by land wholly in the power of Russia, the other by sea equally controlled by the allies, the enemy having dismantled and sunk his ships. The case was therefore reduced to the siege of a great fortress absolutely undisturbed by fears for the communications of the besiegers. In the case of Gibraltar, in 1779–1782, the offense failed through the weakness or imbecility of the allied navies; had these been up to their work, the British fleet could not have thrown in supplies.

The question of waging war in a maritime region beyond the immediate neighborhood of the country so engaged is simply a particular case of general military operations. The case is that of maritime expeditions, in remote waters, where the country may or may not possess already positions useful for the purposes of war; but where in either case it is proposed to make offensive movements, and to possess, or at the least to control, enemy's territory. Even though the leading object of the war be defense, defense is best made by offensive action.

The specific feature which differentiates these operations from others, and imparts to them their peculiar characteristic, is the helplessness while afloat of the army contingent embarked,—its entire dependence for security upon the control of the sea by the navy of its nation. Be it large or small, however efficient in itself in courage, discipline, and skill, it is paralyzed for effective action during the period of transit. The critical nature of this period, together with the subsequent risk to its communications, which depend continuously upon the same control, are the distinguishing elements to be borne constantly in mind while considering this subject.

A few cursory remarks on the leading features of such expeditions in general will first be offered; and then, for the purposes of illustration, two historical cases will be briefly described and discussed.

Having the two fundamental requisites already stated, a reasonably secure home frontier and a navy adequate to dispute control of the sea with the enemy, the next thing is to determine the particular plan of operations best suited to obtain your purpose. This involves the choice of a base, of an objective, and of a line of operations,—three things inherent in every operation of war.

Putting aside, as involving too wide a scope, the question of attacking the enemy elsewhere than upon the maritime region which you wish to control, the ultimate objective there should be that position, line, or district which in its influence upon the general situation may be considered the most important; to use a common expression, the key or keystone. If the particular region aimed at is decidedly nearer to one of your sea frontiers than to the others the base of the plan of operations will be found there, unless prevented by other serious reasons, such as the lack of good harbors or established dockyards. Thus Great Britain, now that Germany has become the threatening naval power, demands another dockyard—preeminently a docking yard—on the North Sea, at Rosyth, additional to the one already possessed at Chatham. This evidently means that, as against the German coast and for operations in the North Sea, the British base is shifted to the North Sea from the Channel, defined by Plymouth and Portsmouth. So Austria and Russia have been pushing territorially towards the Mediterranean, because their other outlets thither, respectively at the head of the Adriatic and from the Black Sea, are too remote, and with communications to the sea too exposed militarily, to be satisfactory as bases of operations.

On the other hand, the United States possesses in its Gulf coast a base line distinctly nearer to the Isthmus and to the western half of the Caribbean than are Norfolk and New York, the two chief naval stations indicated by nature on the Atlantic Coast. Yet it is doubtful whether, with the great increase of size in battleships, and with the difficulties of docking in the Mississippi, the Gulf ports can provide an ultimate base of operations equal to those of the Atlantic; whether they will not rather constitute intermediate advanced ports, valuable as sources of supply because of their nearness, but inadequate to the greater repairs. At the same time, Guantanamo and Key West, in case of operations towards the Isthmus, offer such marked advantages on account of nearness and of mutual support that a secondary provision for docking in them would be highly desirable, and probably expedient. Some military risk must be taken; as is the case, for instance, with the British docks at Gibraltar, which under modern conditions are within range of the Algeciras shore.

The great arsenals define the position of the base line, and in large degree its length. Their local resources, of torpedo vessels and submarines, will tend to protect the seaboard between them, and also a little to each side, from the enemy's operations. The question of moving from the base thus fixed to the objective chosen involves the choice of a line of operations. In the open field of the sea the most direct route is the most natural, and, other things being equal, the best; but many circumstances may influence the decision. Paramount among these is the strength of the navy as compared with that of the enemy,—a strength dependent not only upon aggregate tonnage or weight of metal, but also upon the manner in which those aggregates have been distributed among the various classes of vessels and upon the characteristics of each class in point of armament, armor, speed, and coal endurance. All these qualities are elements in strategic efficiency, sometimes mutually contradictory; and the adjustments of them among themselves may seriously affect strategic calculations. This illustrates that the composition of a national fleet is really a strategic question. The known efficiency of the respective services, and the comparative distance of the belligerent countries from the objective point, which is assumed to be the same for each,—the one to defend, the other to attack,—will also influence the choice of the line of operations, because the length of the lines of communication to be guarded will materially affect the strength of the contestants. It is upon these lines, or belts of sea, that fast cruisers can specially embarrass the operation, compelling the employment of a large proportion of the fleet to check their movements. The shorter and the more numerous these lines, and the farther they pass from an enemy's ports, the greater the task of the enemy and the probable immunity of the lines.

As much of that with which we are about to deal, namely, the advance of a great maritime expedition,—a body of ships of war convoying an army embarked in transports,—may not impossibly seem to some as talking in the air of a thing that never did happen or can happen, bear in mind that it did happen in 1798, under the greatest general of modern times, and that the expedition was pursued by a fleet of about equal size under Nelson, one of the greatest admirals of all time; also, that the questions here raised must have been subjects of careful thought at that time to both Napoleon and Nelson. As to such a thing never happening again, consider the evident future importance of the West Indies and the Caribbean Sea, regarded as approaches to the Isthmus, and through the Isthmus to the Pacific Ocean, in its full extent; also the shortness in time of the distance between any two points in the Caribbean. Remember continually the smallness of that sea, that its length is but one-half that of the Mediterranean, so that great expeditions may well happen there under circumstances peculiarly favorable to such enterprises.

Where a navy is largely preponderant over that of an enemy, such over-sea expeditions by large bodies of troops proceed in security, either perfect or partial. Great Britain during the Napoleonic wars had troops continually afloat, often in large bodies. So did the United States in the Mexican War and the War of Secession. So France in her conquest of Algiers in 1830, and again Great Britain and France during the Crimean War. Security such as existed in these instances leaves little of a military problem; but the case differs when there is an approach towards equality, even though the superiority of one be distinct and emphatic. Vastly superior though the British navy was to the French in Napoleon's time, its tasks were so numerous and onerous that, to quote again, it could not be in force everywhere, and there was always the chance that a hostile division might fall in with an important convoy. Protection localized with the convoy, that is, a body of armed ships in company, was therefore necessary, and the force of these armed ships was proportioned to the importance of the enterprise. It is necessary in this day of ours to remember that the convoys did sail to and fro, and that they were thus protected, fortified, so to say, by armed vessels; for the Blue Water School, or Fleet in Being School, hold that they ought not to sail at all while the enemy's fleet exists in the neighborhood of the line followed. Such convoys of troops were despatched by both belligerents during the War of American Independence, when there was a substantial equality between the opposing navies.

With a navy much superior to that of the enemy,—after allowance made for the length of the line of operations which has to be secured,—it is permissible to strike at once for the coveted objective; the sooner the better. If in so doing you pass by a strategic harbor held by him, capable of sheltering his ships,—a position from which he may with some probability intercept your supplies of coal or ammunition,—this position will require attention to the extent of reducing to manageable proportions the injury possible to it to do.

Thus Jamaica and Santiago lie close to the Windward Passage, which is the direct route from the United States Atlantic ports to the Isthmus; and Cadiz and Gibraltar lie close to the necessary route of all vessels bound from the Atlantic into the Mediterranean. The military characterization of such positions is that "they flank the route." If they harbor ships of war, the route must be protected by force so constituted and so stationed near the port as to check the movements of the ships within. Such a detachment involves exposure to the vessels composing it, in case the enemy has a fleet superior to it anywhere within steaming range. Thus, when the French Brest fleet, in 1799, appeared suddenly in the Mediterranean, the British were divided in several bodies, with more than one of which it might have dealt effectively in succession and in detail. Detachments also entail upon the main body a reduction of strength; but not necessarily, nor always, to such a degree as to arrest progress to the objective. Owing to the variety and importance of British interests, military and commercial, and their wide dispersion, Great Britain in former times found her maritime routes thus flanked in many places by many ports,—by Brest, Rochefort, Ferrol, Cadiz, Toulon, as well as in remoter seas. She met the difficulty by detachments before each, commensurate to those within. The continuousness of this disposition, and the more important military effect it produced in preventing the several hostile divisions from uniting for offensive action, tend to obscure the fact that these various sea positions were thus watched and checked; exactly as a general on shore guards his line of advance against the dangers from a fortress, the position of which threatens his communications in case he is not able to reduce it.

A disposition has been shown lately to cast doubts upon the effect of flanking positions upon lines of communication, as compared to the effect of fortification concentrated upon the objective to which the communications lead. There is no need for such comparison, for no contradiction between the two exists. That Malta can exercise a powerful influence upon the communications of an expedition from a western Mediterranean country acting in Egypt, does not contravene the value of military force in Egypt itself, mobile or fortified. In 1813, Wellington held in Portugal the impregnable fortified position of Torres Vedras, securing against land attack his sea base at Lisbon; yet the puny force of American privateers acting off Cape Finisterre seriously harassed his communications with the British Islands.

If such a point, on or near your line of operations, now become that of communications,—that is, on the line which it was safe for your battle fleet to follow, though not safe for transports,—can be avoided, by making a circuit through waters wholly out of its sweep and nearly safe, your line of communications may be changed. It may be that the character of the port or the known number of its war-shipping will allow the opening of that line from time to time, by convoying transports in large detachments throughout the whole of the more exposed part of the line. Thus the great convoys which at long intervals relieved Gibraltar, 1779–1782, were protected by fleets of battle-ships.

There are two modes by which the supplies of a fleet may be sent forward: one by single supply ships taking their chance, depending upon the routes they follow being controlled by the patrols of their own navy; the other by large convoys under the immediate protection of a body of armed ships. It is probable that both modes will be used; the convoy system being the dependence for the main supply, supplemented by the occasional single vessels. The convoys must be heavily protected, because their sailings should be watched and their destruction attempted by the enemy as one of the regular secondary operations of war; it will therefore be expected to fight a battle for their safety. Single ships must depend upon their speed, upon choosing routes with a view to avoiding danger, and upon the general police of the seas by ships of the cruiser class. Whatever the particular mode adopted, two or more lines of supplies converging toward the objective or toward the position of the fleet are an advantage; more so, perhaps, to a stream of single vessels than to large convoys, as the latter, in any case, must be guarded and so weaken the fighting force. The United States is preparing to fortify the Isthmus, and consequently to garrison the fortifications. It seems evident that, in case of hostilities, it will be expedient that supplies proceed from both the Gulf and the Atlantic, as well as from the Pacific.

A fleet operating some distance from home should not depend upon a single line of supplies. It may be said, generally, that while concentration is the proper disposition for the fighting force, or for preparation for battle, the system of supplies should not be concentrated upon a single line, when avoidable. This statement is, in effect, an application of one of Napoleon's brief, pithy sayings, quoted by Thiers: "The art of war," he said, "consists in the skill to disperse in order to subsist, yet in such manner that you may quickly concentrate in order to fight."

If the reduction of force, by watching an intermediate port, is necessary and also reduces you to an equality with the enemy; or, while leaving you still superior, takes away the chance of overcoming the resistance of the objective before the arrival of adequate relief, then the force should not be divided. Either the wayside port must be taken, or, if you think you can get on with your present supplies until a decisive action has been fought, you may continue on, abandoning your communications for the moment,—cutting loose, as the expression is, from your base and leaving the hostile flanking port nothing on which to work its will. So serious a step, of course, must not be taken without a certainty that the great essential, fuel, will not fail. Without ammunition a ship may run away, human life may be supported on half rations, but without coal a ship can neither fight nor run.

If, as is assumed, the objective is a part of the land, a port or island in the possession of the enemy, the conquest may not immediately and necessarily give the decisive control of the war that a like acquisition on shore may give; because the necessary and limited lines of communication, which often center in such a key to a land region, possession of which conveys absolute mastery, have few exact parallels at sea. This is due to the fact before alluded to and patent to a glance, that, owing to the open character of the ocean, shipping can take many routes to avoid passing near a particular strategic position. A strategic position on land derives much of its importance from the fact that armies are forced to follow certain roads, or incur disproportionate disadvantage by taking others. It is true that there are some near parallels now; as, for instance, Gibraltar. The Russian navy of the Black Sea may be absolutely checked by the possession of the Bosporus or Dardanelles; and it is conceivable that if one great sea power controlled the shores of the North Sea, while another about equal lay along the Channel and Bay of Biscay, the holding of the Straits of Dover by one would seriously embarrass the movements of the other. A canal like Suez is equally such a point.

These instances, however, are exceptional; the power of the keys of maritime regions is like that of the key of a shore theater of war in kind, but falls short of it in degree. The historical instance before cited, the seizure of Egypt and Malta by Bonaparte, precisely illustrates the assertion. The key to the control of the East by the West was in the hands of the French, but they could not use it; nay, it was finally torn from their hands by the naval strength of their foes. What, indeed, is the good of holding the point where roads cross, if you can neither use the roads yourself nor hinder your enemy from using them?

Therefore, if successful in seizing the objective at which you have aimed, by being beforehand with the enemy,—thanks either to better preparation for war, or greater activity of movement, or by being nearer the seat of war,—you cannot think your conquest secure until you have established your naval superiority, and thereby your control of the roads which connect you with home, and also of those the nearness of which to the position you have just taken give it its importance. This superiority *may* be established, it is true, by the conquest itself, the loss depriving the enemy of a necessary naval base and perhaps of a considerable part of his ships; more generally it will arise from your fleet being superior in numbers or quality to his. The same remark is evidently true concerning positions already held before a war breaks out; as, for instance, the Panama Canal Zone and Hawaii, the ultimate retention of which will depend upon the strength of the fleet. It must be remembered that the Monroe Doctrine is not a military force, but only a political pronouncement.

If decisive naval superiority does not exist, you must get ready to fight a battle at sea, upon the results of which will probably depend the final fate of your new gain; as the destruction of the French fleet at the Nile followed Bonaparte's first success, and annulled it. Admiral Togo's signal to his fleet off Tsushima, "The fate of the Empire depends upon this day's work," though primarily an appeal to patriotism, was ultimately and simply a particular application of the general military truth here enunciated. Japan by readiness, skill, and promptitude had projected the national power across sea, forestalling the action of Russia, as Bonaparte that of Great Britain. She had conquered a secure foothold in Korea and Manchuria, and had seized Port Arthur, as the French had Egypt and Malta. The positional keys of the situation were in her hands; but the defeat of Togo's fleet would have annulled all previous successes, as that of Brueys by Nelson did the achievements of Bonaparte. Conversely, Russia also at the same moment had projected Rozhestvensky's fleet, in like manner, close to the position she coveted to attain, accompanying it with convoy and coal destined to further future operations; but within easy range of her point of arrival she had first to fight a battle, in which fleet, coal, and convoy went down to a common destruction.

It may, indeed, very well be that the inevitable battle would have to be fought before instead of after the fall of the position. If the land defenses against which your expedition is moving were weak or in decay, while the enemy's fleet was nearly equal to your own, it would be the duty of the latter to attack you, hampered as you may be in such an expedition with the care of transports and supply ships, and at some distance from the port. Much more will this be the case if the enemy is moving to a home port, from which you aim to debar him, as Togo did Rozhestvensky.

If Nelson, in 1798, could have come up with Bonaparte, he should and would have attacked at sea, for doing which he had made systematic preparation; and, if successful in his assault upon the ships of war, he would have effectually stopped the expedition. In 1759, when large preparations were made in France for the invasion of England, and again in 1795, when it was proposed to send eighteen thousand troops from Toulon for the reconquest of Corsica, it was argued by the French authorities that the fleet should first fight that of the British, because this, being equal to their own, must be got rid of in order to make the passage safely. It must be noted, however, that the inferior skill of the French generally, and of the French admiral in particular, were the principal reasons that the French ministers of marine thus urged on the occasions named.

In such combined expeditions, the question whether the fleet and the convoy should sail together, or the convoy be held till control of the sea is decided, is difficult and disputed. It will be better to offer certain considerations for reflection, rather than to make sweeping dogmatic assertions. It seems evident that much will depend upon the distance of the proposed objective. In contemplated invasions of Great Britain, as by the French in 1759, and by Bonaparte in 1803–1805, or, as if often conjectured, by Germany now, the nearness of the objective, and consequently of the expected naval battle, insures that knowledge of the result will be so speedy that no valuable time need be lost in following up a victory by the transit of the land forces. The enemy will not have opportunity to reconstitute his means of resistance. Consequently there is no adequate reason for exposing the troops beforehand, to share the disaster if the fleet meet defeat instead of victory. Napoleon therefore, in 1805, held his army in leash at Boulogne, awaiting the six hours' control of the Channel which he hoped from the presence of his fleet. As said, the same course had been prescribed in 1759.

But if the object be remote, as Egypt was from France, or as Panama is from the United States or Europe, or Hawaii from every part of the Pacific shores, it may be urgent that landing follow victory quickly, lest the enemy recover his breath in the long interval needed for the troops to come. In such cases, the subsequent landing is one incident, and a very important incident, to following up a victory properly; and it seems entirely congruous to all general principles to say that the means for so following should be at hand. That is, the convoy, the troops, should be with the fleet, in numbers at least equal to the immediate task of seizing a position, and holding it till reinforced. This again is entirely in accord with the methods of crossing a river in face of an enemy. The crossing of the sea is simply a much magnified instance of crossing a stream. It may be that the accompanying troops should be proportioned only to that immediate work, of holding a position till reinforced; but the question of proportion, of numbers, is one of detail chiefly, not of principle, and will be affected by many other details. Naturally, one determining detail would be the desirability of not exposing too many troops to capture upon a defeat of the ships of war.

Again, nearby invasions may be divided into those on a great scale, major operations of war, and those which partake of the character of diversions. The latter may naturally accept risks greater, proportionately to their size, than would be proper in the graver undertakings; because the total hazard is not so great, nor will failure be so disastrous. Chances may be taken with a boat which would be unjustifiable with a ship, and with a ship that would be indiscreet with a fleet. Strategically, the success of a diversion, although it may be eminently contributable to the success of a war, is not vital to it, as is the success of the great main advance. Great expeditions, invasions in force, must have a solid sustained character, which, while demanding rapidity, imposes also a graduated advance, a well-knit system, in which each step presupposes others, and the whole a permanent sustained action, like the invasion of France by Germany in 1870. Diversions, particularly maritime diversions, presuppose rather a momentary action, or at most one the prosecution of which depends upon the turn events take. For them, therefore, all the means of action need to be immediately disposable, in order that the whole may proceed with instantaneous development when the objective point is reached; and this means that the troops must accompany the fleet. The procedure of Great Britain abounds with illustrations of the troops accompanying the fleet. Such were the numerous expeditions against West India Islands, as that against Havana in 1762, and others in the French Revolutionary Wars, and that against the French in Egypt in 1801. These were distant expeditions, on varying scales of size, but in all navy and army sailed together.

Instances from British practice, however, are less illuminative than they would be if the British navy had been less preponderant. More convincing is the action of Hoche, a really great general, in the expedition against Ireland in 1796. His army had been fixed at twenty thousand men, though not so many went. It was to form the backbone of what it was hoped would prove a general Irish revolt; that is, its final procedure would depend upon the turn things took in Ireland. In any event, it would be only a diversion, however influential; but for its success it was imperative that the landing of the troops should follow instantly upon the fleet anchoring, and accordingly they went with it. In 1690 the French contemplated an invasion of Great Britain. It was to be a diversion only, counting on an insurrection in favor of James II. The fleet sailed without the troops. It won a marked victory at Beachy Head, forcing the allied English and Dutch fleets out of the Channel; but the troops were not at hand and the victory was not improved.

As a rule, a major operation of war across sea should not be attempted, unless naval superiority for an adequate period is probable. The reason is that already given, that the main movement of a war should be closely knit by steps linked one with another, which cannot be if the navy cannot command the sea. But promising diversions are permissible even with an inferior navy, the deciding consideration being whether the prospect of gain reasonable overbalances the probable losses from a failure. Where navies approach equality, as in 1690 and during the War of American Independence, the practice generally has been that small bodies of troops are sent out by both sides, without hesitation, under appropriate convoy.

The choice of position in which the defendant fleet would seek to find and fight such expeditions while on their way belongs to the province of strategy. Nelson returning to Europe from the West Indies in 1805 announced to his captains that if he met the allied fleet, of which he was in pursuit, twenty ships to his own twelve, he would fight them; but not, he said, until near Europe, unless they gave him a chance too tempting to be resisted. I do not know that he gave any reason for this purpose as to place. I infer that, against such odds, he would keep in company watching for an opportunity; but that, if none favorable offered, he would fight in any event, because, as he said afterwards, and had said before, "by the time they have beaten me soundly, they will do England no more harm this year." This is a sound strategic motive. The invading force being tied down to one route, or at least to a few, the choice of where to fight will remain in the defendant's hands to a certain extent, depending on his knowledge of the invader's movements.

To illustrate: If Nelson had known where the French fleet was bound in 1798 before the Nile, and could have selected his point of attack, it would have been best to choose it near Egypt; because, in case of reverse to the French, he could destroy their force more entirely at such a distance from France, whereas, if he himself were beaten, the British cause would not be materially worse off because of the distance. The same reasons which lead one party to wish a certain position for fighting, if well founded, should determine the other to avoid fighting there, if possible. Generally, it may be said that the farther from his home base the invader can be made to fight, the better; but this must be tempered by the thought that a small or partial success won close to the objective might not prevent the expedition from landing. Suppose the case of an expedition from Cuba against Santa Lucia. Probably, in the case assumed, the fleet of the defendant, acting offensively, however, as becomes a navy, should try to fall in with the expedition midway and harass it on its course,—a proceeding for which it will have a decided advantage in its freedom of maneuver, having no transports to care for. Such harassment and abiding an opportunity would be precisely the course proposed by Nelson, and, as he also intended, should end in a resolute attack at a point sufficiently far from that of the enemy's destination. The manner of the attack belongs to the province of tactics.

Whether it comes before or after the seizure of the objective, a battle must be fought if a decisive naval superiority does not already exist; and if it does, that superiority must be energetically used to destroy every fragment of the enemy's shipping within reach. The question of naval inferiority need not be discussed; for no such distant expedition as that we are considering, unless it be only a diversion, is justifiable in the face of a superior fleet. When Bonaparte sailed for Egypt in 1798 with thirteen ships-of-the-line, the British had but three in the Mediterranean, although the activity of Nelson enabled him to overtake and pass the expedition after reinforcements reached him. Even so, the British fleet was slightly inferior in force to the French. On the other hand, the French expeditions to Ireland, in the same year and in 1796, failed entirely; through a variety of causes, it is true, but due ultimately to naval inferiority, which forced them to take for the attempt the stormiest time of the year, as most favorable to evasion, and so involved them in disaster. Had a better season been chosen, the effect of British naval command would have been only more direct, and so more apparent. Do not understand this comment to imply condemnation of the particular undertakings. They were projected only as diversions, and appear to have been properly conceived, tested by the standard before advanced; namely, that the reasonable prospect of advantage overbalanced decidedly the probable losses through a failure.

On its way, therefore, such an expedition keeps together as much as possible. It is, for the time, free from care about communications, inasmuch as necessary immediate supplies accompany it. Its anxieties, then, are not strategic, but tactical,—how to protect the fleet of transports, and at the same time maneuver in the face of an enemy, if encountered. Lookout and despatch service is performed by the light cruisers; the heavy ships of the order of battle keep within supporting reach of the admiral and of the convoy.

This sustained concentration of the fighting ships is the primary vital condition. These must not separate, whatever other divisions may occur, or detachments become expedient. The French expedition to Ireland in 1796 might have effected its landing as certainly as did Bonaparte in Egypt two years later, if the military and naval commanders had stuck to the battleships,—had not separated from the fighting force. Bonaparte, perhaps instructed by this experience, kept himself with the admiral in the biggest ship of the line. The convoy of troops for Egypt was collected from several ports while on the way. When that for Civita Vecchia was expected, the French admiral submitted to Bonaparte a written order to detach four ships of the line and three frigates for its protection until it could join the main expedition. Bonaparte wrote on the order, "If, twenty-four hours after this separation, ten English ships of the line are sighted, I shall have only nine instead of thirteen." The admiral had nothing to reply. The incident affords a valuable illustration of the necessity of concentration of thought and purpose; or, to use Napoleon's phrase, of "exclusiveness of purpose" as well as of concentration of force. The admiral's conception was, that by dividing he would protect both the main convoy and the expected detachment. Bonaparte saw that instead of both being protected, both would be exposed to overwhelming disaster; for if the British met the detachment they would be thirteen to four, or, if the main body, thirteen to nine. The detachment would be no safer with four than with nine, and in like manner the main body. The smaller, being for the moment unavoidably separate, must take its chance. The case is absolutely on all-fours with the proposition to divide our fleet between the Atlantic and Pacific, and with the Russian blunder of that character in the late war.

So long as the troops are afloat, the dispositions of the convoying fleet center around their protection, are tactical in character, and governed by the rules applicable to every force on the march when liable to meet the enemy; but when the objective is reached and won, the troops take care of themselves, the tactical dispositions of the fleet for them disappear, and there arises immediately the strategic question of the communications of the army, of the command of the sea, and of the disposition of the fleet so as best to insure these objects. An intermediate hostile port, like Malta, flanking the communications, may then draw upon itself the full, or at least the proportioned, effort of the fleet.

The treatment of our present theme thus far has been by statement of general principles, with only incidental illustration. There will now be cited at some length two historical examples of expeditions such as those under discussion. Separated as the two are by an interval of two thousand years, the lessons which they afford in common illustrate strikingly the permanence of the great general principles of strategy.[2]

Sir Edward Creasy, in his "Fifteen Decisive Battles of the World," ranks among these the defeat of the Athenians before

2. The following to the end of the chapter first appeared as an article entitled "Two Maritime Expeditions," *United Service Magazine*, vol. 129 (October 1893), 1–13.

Syracuse, B.C. 415.[3] Whether the particular claim be good or not, this event certainly has a high value to doubting students of military history, by showing that, under all conditions of material or mechanical development, strategic problems remain the same, though affected by tactical difficulties peculiar to each age.

At the time in question, two centuries before the great strife between Rome and Carthage known as the Punic Wars, Athens had a sea power which was to the world of that day tremendous and overwhelming. It extended over and wielded the resources of the islands of the Ægean Sea, and was strongly based on the coasts of the Dardanelles and the mainland of what we now call Turkey in Europe, whence the trade of Athens was pushed to the Black Sea and to the Crimea, then as to-day the center of a wheat country. During nearly twenty years Athens had been engaged in war with the combined states of the Peloponnesus, now the Morea, at the head of which stood Sparta. Lacking the insular position which has been at once the strength and safety of Great Britain, she had seen her petty land territory in Attica wasted up to the city walls by the far superior armies of the enemy; but she still held out proudly, based upon her great navy and her commercial wealth,—in other words, upon her sea power. By it she mastered and held two distant advanced posts upon the hostile coasts; one at Cape Pylus, now the Bay of Navarino, where the conjoined British, French, and Russian fleets, under Sir Edward Codrington, destroyed the Turkish navy in 1827, the other at Naupactus, at the entrance of the Gulf of Corinth. Both were valuable strategic points for making raids into the enemy's territory, and for intercepting the corn trade from Sicily, which island was peopled by Greeks of a race kindred to Sparta. Besides these, the island of Corcyra, now

3. See map on page 206.

Corfu, was in strict alliance with Athens; and as in those early days the course of galleys bound from Greece to Sicily was to coast to Corfu, then across to the Iapygian Promontory, now Cape Santa Maria di Leuca, and thence to follow the Italian coast, the strategic worth of the island to those who controlled it is easily seen. It therefore was chosen as the point of assembly for the transports; but the great bond knitting together all the elements of strength was the Athenian navy.

This was the situation when, towards the year B.C. 413, Athens determined upon the conquest of Syracuse, the chief city of Sicily, as a prelude to the subjugation of the whole island. Of the many motives leading to this serious step, involving a reach over sea much exceeding any previous attempt, we are here concerned only with the military; and but incidentally with them, as our attention is called mainly to the expedition proper and not to the whole war.

The military reasons for attacking Sicily were, first, the fear that its Greek cities, being mainly colonies of a race antagonistic to Athens, would join with their fleets in the war then raging. If this fear was well founded, sound military policy justified, nay, demanded, an attack in force upon them before they were ready; for, if a junction were effected, they would seriously imperil the control of the sea, upon which the safety of Athens depended. The condition much resembled that of 1807, when the British Government seized the Danish fleet at Copenhagen, it having become known that Napoleon, in understanding with the Czar, proposed to compel the coöperation of the Danish navy in his general naval policy. The second reason was that Sicily, as fruitful then as in all ages, supplied the enemy with wheat, even as Athens drew her grain supplies from the Black Sea. The two together justified the undertaking from a military standpoint, if there was strength enough to succeed in it and to hold the chief seaports; and after weighing the blunders of the Athenian commander-in-chief as well as I can with an unavoidable lack of experience of the sea difficulties of that day, I am persuaded the sea power of Athens was equal to the task.

So much for the general military policy. Let us now examine the conditions of the particular field in which this expedition was to act.

Athens was on the side of Greece farthest from Sicily. The Peloponnesus, the territory of her opponents, lay between it and her. She held the islands Cythera (Cerigo) off the south of the peninsula and Corcyra on the west, while by her sea power she controlled the other Ionian Islands and occupied the seaports Naupactus and Pylus. Along the south coast of Italy, which fleets would usually follow, every city was hostile or unfriendly, until on the Straits of Messina Rhegium was reached; but Tarentum and Locri, at the two ends of this line,—the one at the heel, the other at the toe of the boot of Italy,—were strenuously inimical. Messene, on the Straits of Messina, had passed from one party to the other, but was now held against Athens. Then came, on the east coast of Sicily, three friendly cities, beyond which Syracuse, the first objective, was reached. The misfortunes of the expedition never allowed a thought of a farther step.

The choice of Syracuse as the objective was accurate. It was the front and center, the foundation and keystone, of the whole system of danger to Athens from the western colonies. To strike at it direct was right, if the sea strength of Athens was as great as the event showed. In so doing, the expedition passed by the hostile strategic points capable of sheltering an enemy's fleet; but the leader had reason to think that, if unaided, they would not dare to act against him. He was right; their jealousy of Athens refused help, beyond water and permission to anchor, but no seacoast city dared to lift a hand against the power of the sea. The Athenians thus cut loose from their base, having force amply sufficient to crush Syracuse before help could come, depending, with reason, upon their control of the sea daunting the enemy cities near their communications. In fact, these hostile ports, except Tarentum, finally allowed the besiegers of Syracuse to be supplied from their markets, and thus became to them new bases of support, of resources.

This was what did happen. What might have happened—with the respectable though inferior fleet possessed by Syracuse, and under the strategic conditions, added to the tactical embarrassments inseparable from an expedition composed of ships of war and transports—is admirably set forth in a speech made by a Syracusan before the popular assembly of the city. This man, named Hermocrates, proposed to make active use of the strategic position of Tarentum, flanking the Athenian line of operations, by sending thither a fleet which would either deter the enemy by threatening his communications, or, if he persisted, would act offensively against them and the fleet as opportunity arose,—Nelson's opportunity, too tempting to be resisted. The speech of Hermocrates was as follows:

"There is one point more which, in my opinion, is more critical and important than all the rest; and although, inured as you are to domestic indolence, it may perhaps not gain your ready approbation, I shall, however, boldly recommend it. If all of us in general who are inhabitants of Sicily, or at least if only we Syracusans, with what other people we can get to assist us, would put out instantly to sea with all the ships we have in readiness; and, victualled but for the space of two months, if we would then give these Athenians the meeting either at Tarentum or at Cape Iapygia, and there convince them that before they enter the lists of war for the conquest of Sicily they must fight for their passage across the Ionian Sea, we would strike them with the utmost terror. We would infinitely perplex them with the thought that from a friendly port we sally forth to guard our outworks, for Tarentum will readily receive us; while they have a long tract of sea to pass with their cumbersome train, and must find it hard, through so long a voyage, to be always steering in the regular order. As their course must thus be slow, and must advance only in exact conformity to order, we shall have a thousand opportunities to attack them. If, again, they clear their ships for action and in a body bear down expeditiously upon us, they must ply hard at their oars, and, when spent with toil, we can fall upon them;[4] or, in case fighting be not advisable, we have it always in our power to retire into the harbor of Tarentum.

"Thus, if the Athenians, in constant expectation of being fought with at sea, make their passage with but a small portion of their stores, they will be reduced to a great distress upon coasts which will afford them no supply. Should they choose to continue in their station," that is, to remain in Corfu, "they must be infallibly blocked up in it; should they venture a passage, they must unavoidably leave their tenders and storeships behind," because of the tactical embarrassment in the day of battle, "and as they have no assurance of a hearty reception from the cities of the coasts, must be terribly dismayed," for their communications. "It is my firm opinion that, amidst the great perplexity of thought which must result from these obstructions, they will never presume to set sail from Corcyra; or, at least, while they are agitating the forms

4. In the tactical system of the Athenians, who were the most expert of seamen, dependence was placed upon their superior skill enabling them to charge the sides of an enemy's ship with the bow of their own,—the strongest part of a vessel to the weakest. For this maneuvering the rowers needed to be in full strength,—fresh. The Syracusans reinforced the bows of their galleys abnormally, to meet bow to bow, like two eggs testing strength. Their inferior skill could not insure this particular collision if the Athenian rowers, with their lighter handling vessels, were fresh. The Romans would not be affected by the fatigue of rowers, if they could reach the enemy; for then they made fast, and the fighting men, who were fresh, fell on. All these particulars are tactical; there is in each of them just as real tactical resource as in the American navy replacing eighteen pounders by twenty-fours in 1800–1812.

of procedure and sending lookout vessels to discover our numbers and positions, the season of the year will be protracted to winter."

A further detail, affecting naval operations in the field of strategy as well as of tactics, is evidently deducible from this speech; namely, if the ancient fleets proposed to themselves to keep the sea for some time, as in this instance to make a straight course from Corcyra to Syracuse, they were forced to carry such a weight of provisions and water as brought them down in the water and rendered them slow and difficult in maneuvering. In other words, the strategic consideration of the route to be followed, whether as shorter or with reference to the friendly disposition of the coast skirted, involved also the tactical question of the efficiency of their fleets in the all-important point of speed and turning power. The same thought applies to overloading with coal when within tactical reach of the enemy, as Rozhestvensky did. It is also worth remarking that the excessive labor of the oar necessitated for the oarsmen an ample supply of nourishment, and especially of fluid in warm weather.

The salient and decisive features, however, in the plan of Hermocrates were: (1) The recognition of the strategic value of Tarentum, flanking the route the enemy must take, as do Jamaica, Gibraltar, Malta; and (2) the use to which he proposed to put this, by rapidly mobilizing the Syracusan or Sicilian fleets, and massing them on the flank of the Athenian line of advance in a position secured from attack. Here was, first, a threat the enemy could scarcely venture to disregard; and, secondly, the preparation of the inferior navy for offensive action at a moment's notice, and upon the weakest yet most vital link in the enemy's scheme of operations.

In the proposition of Hermocrates, then, we have a true and fruitful strategic thought, with the modification due to tactical conditions, put forth two thousand years ago by a man who never heard the words "strategy" or "tactics" technically used, nor tried to formulate their laws. If, however, any one is disposed to infer, from the accurate insight of this untaught genius, the uselessness of studying war as an art, he may be quickly set right by the coldness and insult with which the speech was received, and the vainglorious appeal to national self-sufficiency made by the opposition orator; a Grecian anticipation of "buncombe," which can be read in Thucydides. The advice was rejected with contumely, the result being the unopposed progress of the Athenians, and the consequent siege, suffering, and narrow escape of Syracuse, with the change of attitude before mentioned in her friends, the Italian-Greek cities. However many and mixed the motives of the Syracusan assembly, a knowledge of the principles of war might have given the true policy a chance; averting the ruin which nearly overtook the city, and would have overwhelmed it but for the imbecility of the Athenian general.

It should be added, however, that although the scheme of Hermocrates was not only sound, but the very best fitted to the conditions, it would not assure the certainty of Syracuse's safety, because she was much the inferior power. It was the most skilful thing to do; it secured the most numerous chances; but if the Athenian skill had been equal, the stronger must in the end prevail. Indeed, the transfer of the Syracusan fleet to Tarentum, as conceived by Hermocrates, illustrates aptly both the power and the limitations of a "fleet in being," of which we have heard and still hear so much. Tarentum would have fixed upon itself the Athenian attack, just because the hostile navy was there; as Port Arthur fixed the attention of the Japanese, and Santiago that of the United States in 1898. Syracuse would have been saved by the fleet, at least until Tarentum fell. The momentary safety of Syracuse would illustrate the influence of a "fleet in being"; its subjugation after the fall of Tarentum would show the limitations of such a fleet, which, by definition, is inferior.

This episode in the Peloponnesian War, for in result it

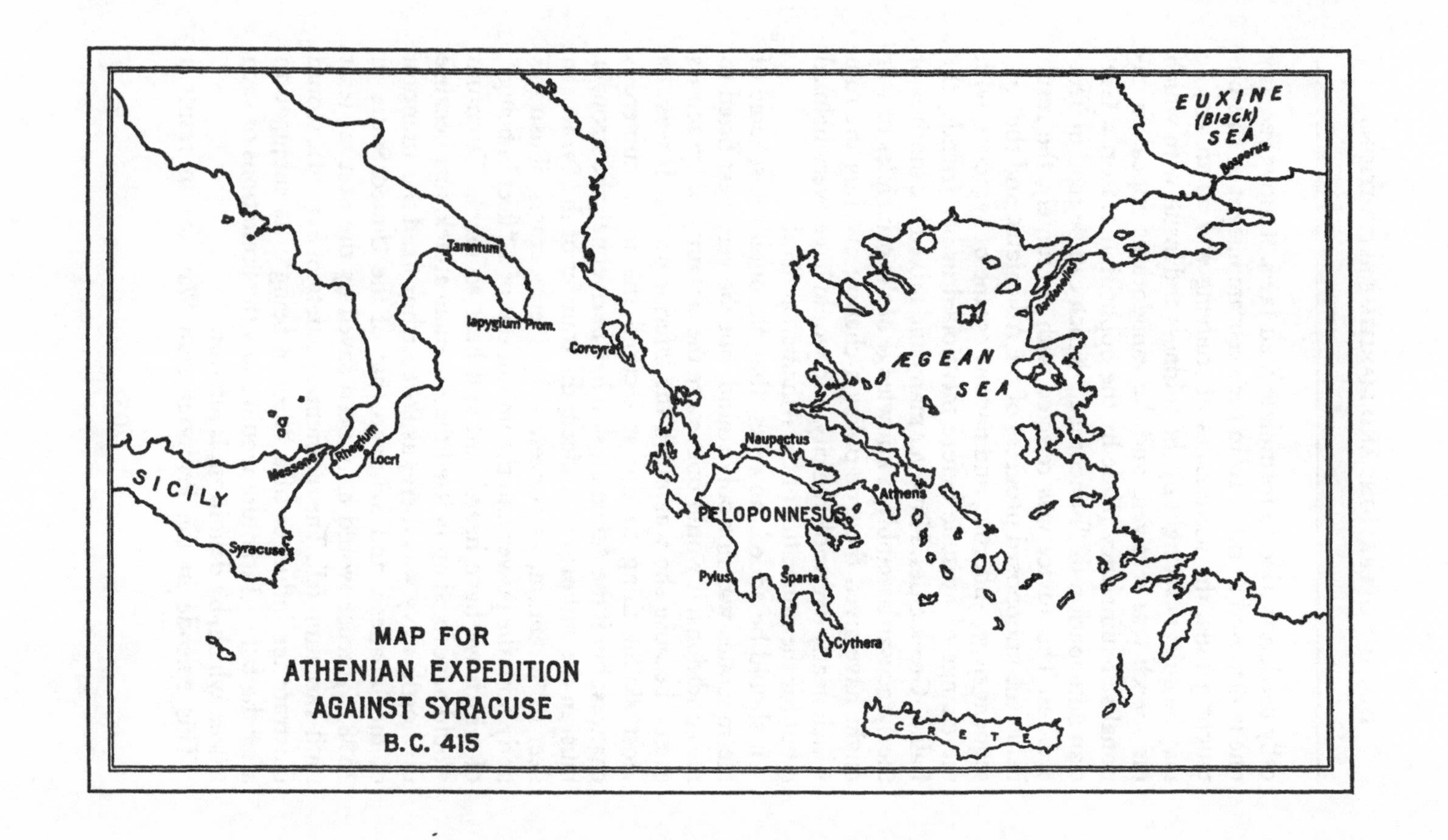
EUXINE (Black) SEA
Bosporus
Dardanelles
ÆGEAN SEA
Tarentum
Iapygium Prom.
Corcyra
Naupactus
Athens
PELOPONNESUS
Pylus
Sparta
Cythera
CRETE
Messene
Rhegium
Locri
SICILY
Syracuse
MAP FOR
ATHENIAN EXPEDITION
AGAINST SYRACUSE
B.C. 415

proved to be no more, gives us all the conditions of a distant maritime expedition in any age. We have the home base, Athens; the advanced intermediate bases at Corcyra and other points, which played for Athens the part that Gibraltar, Malta, and foreign coaling stations have done and still do for Great Britain; the objective, Syracuse; the neutral, doubtful, or hostile country to be passed, across the Ionian Sea or along the coasts of Italy; the enemy's advanced post in Tarentum and sister cities; the greater naval power in Athens; the smaller but still respectable fleet of Syracuse; the difficulty of communications; the tactical embarrassment of a train of supply ships; the tactical difficulty of ships deeply laden for a long voyage, which exists in a degree to-day; the tactical difficulty of the fatigue of rowers, which has disappeared; the wisdom of meeting the enemy half way and harassing his progress; the danger of awaiting him at home on the defensive; the perception of a navy's true sphere, the offensive. All these broad outlines, with many lesser details, are to be found in this Athenian expedition, and most of them involve principles of present application. In fact, put this early galley expedition under a microscope and there is seen realized the essential leading features of every maritime invasion.

The attempt of the Athenians, though overwhelmingly disastrous in this issue, was justified, because they were by far the superior naval power, and therefore had the probability of operating over a controlled sea.

The same cannot be said for Bonaparte's Egyptian expedition of 1798.[5] Without attempting to analyze the mingled motives which determined the action of the French general, it is probable that he was swayed largely by his disposition to trust to the chances of war, as he so often did both before and after, and, for long, not in vain. Also, when the expedition left Toulon, there were but three British ships-of-the-line in the

5. See map on pages 240 and 241.

Mediterranean, while even of their presence he may well have been ignorant. There was, therefore, a very reasonable preponderance of chances that the landing could be effected before an interrupting force could come up. Once on shore he was not wholly unjustified in relying for further progress upon the resources of the conquered country and upon his own unsurpassed powers for war and organization; and there seems reason to believe that, by adopting for the fleet a course analogous to that recommended by Hermocrates, he might greatly have increased the perplexities of the British admiral, and by so far his own chances of success.

His foot firmly planted in Egypt, Bonaparte, having compassed his first objective and so far accomplished his offensive purpose, necessarily passed, as to Great Britain, to the defensive with an inferior navy. This was precisely the position of Syracuse in relation to Athens, and the question may be considered, "What use should he have made of his fleet?" Having supreme command of it as well as of the army, this care was constantly in his mind. There were many considerations, political and administrative, that must justly have influenced him; but from the purely military point of view his decision appears to have been about the worst possible.

Good communication with home was the one thing necessary to his final success; nay, to the very existence of the French army in Egypt. There was no doubt of its ability to subdue Egyptian opposition; but it was bound to suffer losses by battle and disease, and if it advanced, as it must, there was further loss by unavoidable dissemination of forces. The numbers needed frequent reinforcement. Certain supplies also must come from home, such as ammunition of all kinds and equipments for war, not to speak of the moral effect upon the army of finding itself cut off from any probable hope of returning to France. There was, too, the possibility that the fleet, under favorable circumstances, might coöperate with the army; as, indeed, frigates did a few months later in the Syrian expedition.

The danger that threatened all this was the British fleet. No port nor number of ports along the line—as, for instance, Malta, which the French held—could keep communications open if that fleet were left untrammeled in its movements. It was now known to be in number approaching that of the French, although the French admiral continued in a state of blissful confidence about his power to resist it.

Criticism, always wise after an event, condemns as visionary Bonaparte's attempt upon Egypt, as it also has that of the Athenians against Syracuse. Having paid some attention to the matter, my own opinion is that, though the probabilities were rather against his success than for it, there were chances enough in his favor to justify the attempt. Much military criticism consists simply in condemning risks which have resulted in failure. One of the first things a student of war needs to lay to heart is Napoleon's saying, "War cannot be made without running risks." The exaggerated argument about the "fleet in being" and its deterrent effect upon the enemy is, in effect, assuming that war can and will be made only without risk. What a risk was run by General Grant when he went below Vicksburg, against which Sherman remonstrated so earnestly, or by Farragut when he passed the forts below New Orleans, leaving them in control of the river behind him.

The orders of Bonaparte were clear and precise, that the ships of war should be taken into the old port of Alexandria, if there was water enough on the bar; if not, the admiral was to go to Corfu, then in possession of the French, or to Toulon. These orders looked first to the safety of the fleet; and next to keeping it, if possible, under Bonaparte's own control. The retention in Alexandria was open to two objections: the first, tactical in character, was that the fleet, though perfectly safe, could be easily blockaded there, and could with difficulty come out and form in the face of an active enemy; while, secondly, there was the strategic inconvenience that its presence there would draw the British fleet to the precise point where transport ships and supplies from France must converge. The French navy, in taking this position, would give up entirely its special properties, mobility and the offensive,—which Hermocrates was so careful to insure,—and for the purpose of keeping open communications with home would be as useless as it became after the Battle of the Nile. It is truthfully remarked by a French naval officer that, with the difficulty of exit, a fleet in Alexandria could be checked by an inferior force, which could fall on the head of the column as it came out of the narrow entrance.

The admiral disobeyed these orders, and for the worse. He anchored near Alexandria in an open roadstead, presenting to an enemy's attack no difficulties except hydrographic; and his dispositions to strengthen the defense were slothful and faulty. The question of engaging the enemy under way or at anchor was discussed in a council of war, where it was decided to await them at anchor; and the line of anchorage was established with that view. This decision, which, it will be noted, was tactical, not strategic, was as unfaithful to the true rôle of the navy as were the orders of Bonaparte for its strategic disposition. Tactically, the fleet was devoted by its commander to a passive defensive, giving up its power of motion, of maneuver, and of attack. Strategically, Bonaparte, in this case, was relying upon the deterrent effect of the "fleet in being" upon Turkey.

It is, however, generally admitted that a strategic fault is more far-reaching than one of tactics, and that a tactical success will fail of producing its full effect if the strategic dispositions have been bad. We may therefore fasten our attention upon the strategic error. Had the result of the Battle of the Nile been favorable to the French, and the fleet been afterwards withdrawn into Alexandria according to Bonaparte's orders, there would have been no positive gain to the Egyptian expedition. In the supposed case, the French fleet would have gained an advantage over the British navy by inflicting upon an isolated detachment a certain loss, perhaps even a disabling loss; but the purpose of Bonaparte to keep it under his own hand at Alexandria would have rendered the success futile, because, wherever the French fleet was, it drew an equivalent British detachment with the force of a magnet, and before Alexandria such a detachment was in the most favorable position to intercept supplies coming from France.

Granting that the strategic disposition proposed was faulty, to what use should the fleet be put?

It is in a case of this kind that the helpfulness of principles clearly and strongly held is felt. It is a very narrow reading of the word "principle" to confine it to moral action. Sound military principle is as useful to military conduct as moral principle is to integrity of life. At the same time it must be conceded that the application of a principle to a particular case is often difficult, in war or in morals.

If the principle is accepted upon which Hermocrates acted, perhaps unconsciously, that the part of the navy in a defensive operation is to stand ready for immediate offensive action, and to threaten it, it is seen at once that a provision which looked only to its safety, while neutralizing its power of movement, such as shutting it up in Alexandria, was probably erroneous. If found there by the British fleet, it was caught in a vise.

If not in Alexandria, then where? Bonaparte named two alternative ports, Corfu and Toulon. As regards these, (1) the fleet could not have been shut up in either as easily as in Alexandria; (2) a larger force would have been needed to keep it in; and (3) a fleet employed in doing so would have been less able to stop the French communications with Egypt. At Toulon, it is true, the British fleet would have been at the strategic point whence supplies would in most cases start; but it is easier to get out of a blockaded port than into it, while, as for shutting up a large fleet, it was difficult for an enemy to keep close to Toulon in winter.

Corfu, however, Bonaparte's third alternative, offered very distinct and decisive advantages as a position over Toulon. A British fleet watching off that island would have been far removed from the direct route between Toulon and Egypt,—over three hundred miles, two days' sail at least, not to speak of the difficulty of receiving intelligence. The office of the navy as accessory to a defensive position, such as was that of the French in Egypt relatively to Great Britain, is to keep open communications by acting, or threatening to act, upon the offensive. This can only be done through its power of movement on the open sea, and by assuming an initiative suited to its strength whenever opportunity offers; for the initiative is the privilege of the offense. Keeping communications open on a given line means either drawing or driving the enemy off it. If not strong enough to drive him off, then diversion must be attempted,—by threatening his interests elsewhere and in as many quarters as may be, seeking to mislead him continually by all the wiles known to warfare. As with war in general, this is a business of positions, and consequently the chief station of a force destined to exert such influence is a matter of prime importance. Corfu, though not without its drawbacks, fulfilled the conditions better than any other port then under French control, because the mere presence there of a French squadron detained the enemy's fleet so far from the vital line of communications. Malta, on the contrary, like Toulon, would have fixed the enemy on the very road which it was desirable to keep open.

Drawing away, or diverting, assumes that strength is inferior to the enemy's, as was then the case with the French navy, as a whole, relatively to the British. If, however, though weaker in the aggregate, you are stronger than a particular detachment encountered, it should be attacked at once; before reinforced. Such weak detachments commonly result from an enemy's fears for his exposed points; as, for instance, Brueys, as mentioned already, wished to make a detachment from his main fleet, because he feared for the safety of the troops from Civita Vecchia. Therefore the aim of the weaker party should be to keep the sea as much as possible; on no account to separate his battleships, but to hold them together, seeking by mobility, by frequent appearances, which unaided rumor always multiplies, to arouse the enemy's anxieties in many directions, so as to induce him to send off detachments; in brief, to occasion what Daveluy calls a "displacement of forces" unfavorable to the opponent. If he make this mistake, either the individual detachments will be attacked one by one, or the main body, if unduly weakened.

These movements are all of a strategic character or aim. If, as a result of them, a collision is effected with a part of the enemy's forces,—say in the proportion to your own of two to three,—a strategic advantage will have been achieved. In the action that should then follow the aim will be to increase this advantage of numbers by concentrating as two to one, or at least in some degree of superiority on part of the enemy's ships. This, however, belongs to the province of tactics, or, more accurately, grand tactics.

Let us now apply these principles to the probabilities in the case before us. The facts were as follows: Nelson first appeared before Alexandria on June 28th, 1798, three days before Bonaparte. Not finding the French fleet there, he supposed he had mistaken its destination and hurried back to the Straits of Messina, anxious about Naples and Sicily. If during the month of July, while he was thus away, the French fleet had sailed as is here advocated, Nelson, on returning, would have found the following state of things: The French army ashore with its supplies out of reach; the transports and frigates in the port of Alexandria, equally inaccessible; and the enemy's fleet gone, on what errand of mischief he could not tell. To remain there with his whole force would be useless. To follow with his whole force would be correct by all principles; but if he did so, he left Alexandria open. The temptation to leave a blockading detachment, say two ships (he had then no frigates), would be very great.

Alexandria, however, was not the only port in the power of France and connected with the whole system of her Mediterranean control. At this moment she held firmly both Toulon and Malta; and Corfu with a grip which later on resisted for a certain time the attack of the combined Turkish and Russian squadrons. It was a matter of importance to her enemies that she should not strengthen herself in Malta and Corfu. All these four points, Toulon, Malta, Alexandria, and Corfu, therefore claimed the attention of Great Britain; yet all could not be effectually watched, so as to break up the communications with Egypt, without a dividing of Nelson's fleet which would have made each fragment hopelessly weak. This was exemplified a year later, in 1799, when the incursion of the Brest fleet into the Mediterranean, before alluded to, found the British thus divided.[6] To the power of distraction thus in the hands of France is to be added the unprotected condition of England's allies or friends in Naples, Sicily, and Sardinia; all of which were open to attack from the sea. In short, the French fleet, until cornered, was facilitated in any operations it might undertake by the possession of several secure ports of refuge, well spaced and situated; and had besides large power to inflict injury, and to exact contributions and supplies from states committed to the side of Great Britain. Only a half-dozen years earlier the squadron of Latouche Tréville had compelled such acquiescence from the kingdom of Naples at the cannon's mouth.

It will be seen, therefore, that so long as the French had in

6. See *Influence of Sea Power upon French Revolution and Empire*, vol. i, p. 304.

the Mediterranean, ready for action, a compact body of thirteen serviceable ships-of-the-line, the number engaged at the Nile, a force slightly superior to Nelson's own, besides frigates, Nelson's fleet had several different objects, all of importance, demanding his attention. There were the four hostile ports just mentioned, the enemy's communications to Egypt, and the protection of Great Britain's allies. Besides all this there was the additional object, the French fleet. There can be little doubt that the genius of Nelson would have led him straight for the enemy's fleet, the true key to such a strategic situation. But not every admiral is a Nelson; and even he could not have stopped the communications effectually before he had found and beaten the fleet. After the battle of the Nile Nelson's ships scattered: some to Naples, some to Malta, some to Gibraltar, some before Alexandria. Such a dispersal shows sufficiently the exigencies of the general situation, but it would have been simply insane under the present supposition of thirteen French ships of the line compact in Corfu. In the positions taken afterwards, each of the small British detachments named did good service, and in perfect safety, having the enemy's fleet off their minds.

But if, instead of being at sea in the Mediterranean or in port at Corfu, the French fleet had been safely moored in the harbor of Alexandria, Nelson's task would have been simpler. The blockade of that port would have settled the communications and left him perfectly easy as to the fate of Naples and Sicily. Toulon lost most of its importance with the British fleet thus placed across the path of any reinforcements it might send; and Nelson from his fourteen ships could safely have spared two, if not three, for the blockade of Malta, which, however, could have been maintained by frigates. It will be remembered that in Alexandria the French fleet labored under the tactical disadvantage of having to come out by a passage narrow, so far as sufficiency of water was concerned, while the outsiders, during the summer, had a fair wind with which to approach. In other words, the French fleet would have to come out in column, in face of a resolute enemy able to deploy his vessels across its head and to fall upon the leading ships in detail.

It may be objected that the quality of the French fleet was so inferior that the suggested use of it as a cruising squadron could end only in disaster. There is some truth in such a statement; and there were also political and administrative reasons, as already admitted. If engaged in a criticism upon the management of the French navy under Bonaparte in its entirety, full weight would have to be allowed to these considerations. Using the case, however, simply as an illustration of strategy, as is here done, it is permissible to disregard them; to assume, as must be done in an abstract military problem, a practical equality, where numerical equality is found. None but a hopeless doctrinaire would deny that circumstances powerfully modify the application of the most solid general principles; yet principles are elicited only by eliminating from a number of cases those conditions which are peculiar to each, so that the truth common to all becomes clear.

As regards the particular effect upon Nelson of such a use of the French fleet as here advocated, it must be remembered by us, who now see things with the facility of what our American slang calls "hind-sight," that if ever Nelson lost his head, was "rattled," it was when he at his first visit found the French not in Alexandria; and after all was over, he spoke most dejectedly of the state of his health, induced by "the fever of anxiety" through which he had passed. We now may dismiss lightly the contingencies here suggested; that great seaman felt their pressure and knew their weight. Besides, it may be said that if Nelson had missed the French twice more, or a few weeks longer, he might have lost his command, so great was the popular clamor over his first failure; and there was scarcely another British admiral at that time fitted to deal decisively with an equal enemy. Napier estimates the presence of Napoleon on a battlefield as equal to thirty thousand men; and it is no exaggeration to say that Nelson, for thorough dealing with an enemy's fleet, was equal to a reinforcement of three ships-of-the-line. The mere success of the French fleet in dodging pursuit and raising alarm might have cost Great Britain her most efficient sea-commander.

It may be further interesting to note that the course recommended by Hermocrates, and here suggested for the French fleet, is identical in principle with certain well-known instances in land warfare. When the allied Austrians, British, and Dutch were falling back through Belgium before the victorious advance of the French in 1794, upon reaching a certain point they separated; the Anglo-Dutch retreating upon Holland, with the vain thought of covering it by direct interposition, while the Austrians took the road to Germany. Jomini ("Guerres de la République") shows that if, instead of this folly of separating, they had massed their entire force in a well-chosen position covering their communications with Germany, and to one side of the line leading to Holland, which was the objective of the French, the latter could not dare to pass them, leaving their communications open to attack. They must have stopped and fought a pitched battle upon ground of their enemy's own choosing before they could touch Holland.

This again, in 1800, was the principle dictating Bonaparte's order to Masséna to throw a heavy garrison into Genoa. The Austrians could not pass by it, could not advance in full force along the Riviera against southern France, while that garrison flanked their line of communication. They were compelled to mask the place with a large detachment, the withdrawal of which from the main body vitally affected the campaign. This again, in 1808, was the real significance of Sir John Moore's famous advance from Portugal upon Sahagun in Spain.[7] The

7. See map on page 226.

threat to the French communications arrested Napoleon's advance, postponed the imminent reduction of Spain, gave time for Austria to ripen her preparations, and entailed upon the emperor, in place of a rapid conquest, the protracted wasting Peninsular War, with its decisive ultimate effects upon his fortunes. Napier shrewdly says that his own history might never have been written if Moore had not made the move he did.

But let it be remembered always that the strength of such dispositions lies not in the inanimate fortresses so much as in the living power of the men, troops or seamen, whose purposes they subserve. To use Napoleon's phrase, "War is a business of positions," but not so much on account of the positions themselves as of the men who utilize them. Of this, the uselessness of Malta to the French in 1798–1800 is a conspicuous example. It flanked the line of communications from the West to the Levant; but, there being no fleet in the port, the position was useless, except as engaging the attention of a small British blockading force.

CHAPTER VII

OPERATIONS OF WAR

THE last lecture began with some cursory remarks on the subject of maritime expeditions in general, and then was illustrated by two special historical cases of such expeditions.

The length to which this illustrative digression went makes it necessary to recall that the question leading to it was this: What is the true strategic use to be made of the naval force when the key to a maritime region, or any advanced position of decisive importance in such a region, has been won by a combined expedition? The answer given was that, when such a success had been won, the particular expedition, having next to secure and preserve that which had been gained, passes from the offensive, with which it started, to the defensive, and that the true part for the navy to bear in such a defensive is the offensive-defensive. When the first objective is possessed, the navy, heretofore tied to the rest of the expedition, is released, the army assumes the defense of the conquest, or the further prosecution of the conquest, and the fleet takes charge of the communications, and so of its own element, the sea. But it can fulfill such a charge only by either driving or drawing the enemy's sea force away from the region in dispute or from the critical point of the campaign. If stronger, it will seek, and if possible compel, a battle; if

Chapter X, *Naval Strategy* (Boston, 1911), 243–301.

weaker, it will try to draw the enemy away and to divide his forces by threatening other strategic points or vital interests. It should be noted that this is precisely the function of a navy in relation to the defense of the home coast-line, if the nation be reduced to a naval defensive for the war.

The Emperor Napoleon in the early part of 1812 discussed a somewhat similar situation in land warfare, in a letter of instructions sent to Marshal Marmont, commanding in Spain that part of the French forces which lay about Salamanca and in face of the fortress Ciudad Rodrigo.[1] This, strong though it was, the British under Wellington had recently captured by an operation which in its swiftness resembled rather a *coup de main* than a siege. Western Spain, bordering Portugal, was a region which the French had seized and from which the British sought to dislodge them; for the French at this period were reduced to a defensive attitude in Spain, owing to the approaching war with Russia, which had led the emperor to concentrate the most and the best of the troops at his disposal upon the great Russian expedition.

The borderland of Spain and Portugal thus corresponded to our maritime region: belonging to neither party, occupied by one, sought by the other. There were in it two principal fortresses, answering to fortified seaports,—Ciudad Rodrigo in the north and Badajoz in the south, both of which had been occupied by the French. Upon the tenure of these depended control of the region. By a rapid movement Ciudad Rodrigo had been taken from them, as just said. Badajoz had been threatened also; it was one of the two keys to the frontier, and now the only one remaining to France. Before the capture of Ciudad Rodrigo it had been besieged in form for some time by Wellington, taking advantage of the weakened condition of the French army in consequence of a disastrous retreat from Portugal the year before, 1811.

1. See map on page 226.

Marmont had succeeded to the command while Badajoz was besieged, and before Ciudad Rodrigo fell. Collecting his army in the field, his mobile force, corresponding to the navy in sea warfare, he marched towards Badajoz. Wellington, unable to sustain both siege and battle, raised the siege, retired into Portugal, and thence moved north of the Tagus to Almeida, there and thus confronting Ciudad Rodrigo, and watching. Marmont also returned to the north, to Salamanca, and had he remained expectant, on a concentrated defensive, ready to act offensively if need be, the presence of his force would have fixed that of Wellington; but he conceived the idea of sending help to a brother marshal, Suchet, then besieging Valencia in the east. As he moved south towards the Tagus, the British thought at first that he intended to invade southern Portugal from Badajoz; but as soon as he detached his five thousand men to the eastward Wellington recognized that south Portugal was not threatened. He saw also that north Spain had been stripped momentarily of effective French force, for Marmont had sent with the detachment a large part of his artillery and cavalry. Then the British swooped down upon Ciudad Rodrigo, carried and garrisoned it before the field army—the navy—could come to its support. The chance was close, and therefore the place was stormed before the time was ripe by the rules of the engineer's art. "Ciudad Rodrigo *must* be stormed to-night," read Wellington's orders; and his army understood the appeal to its courage to indicate the near return of the French relieving force.

It is to the general situation hence resulting that Napoleon's letter of instructions applies. He writes to Marmont,

> Your army being now strong, equipped again with siege artillery, and restored in *morale* as well as numbers, it is no longer necessary, in order to protect Badajoz, that you march upon it. Keep your army in divisions around Salamanca, spread out sufficiently for ready subsistence, but at such distances that

> all can unite in two marches; that is, in two days. Make all your dispositions such that it may be understood you are preparing to take the offensive, and keep up a continual demonstration by outpost engagements. In such an attitude you are master of all the British movements. If Wellington undertakes to march upon Badajoz, let him go; concentrate your troops at once, march straight upon Almeida, and you may be sure he will quickly return to encounter you. But he understands his business too well to commit such a fault.

You have here the French army in the field, the mobile force corresponding to a navy, protecting its acquired post, Badajoz, by detaining the enemy's field army; that is, by drawing it away, or keeping it drawn away, from the position which it had shown its wish to capture. In the first instance, Badajoz had been saved by Marmont's approach, *driving* away Wellington. In the second, Ciudad Rodrigo was lost by misdirection and dissemination of Marmont's force; the British army in the field snatched away an important position. In the third, Badajoz is protected, not by direct action, but by the indirect effect of a sustained diversion menacing interests which the British could not afford to neglect.

The same method of diversion was projected on a gigantic scale by the Emperor Napoleon in 1804 and 1805, when he wished to draw away a large proportion of the British fleet from Europe, and in their absence to concentrate his own navy in the English Channel, to cover the descent upon Great Britain. It will be remembered that France during this period was on the defensive as regards its coast-line. By his formulated plan the fleets from Toulon, Rochefort, and Brest were all to escape from their ports, to meet in the West Indies, and to return in a body to the Channel. It was expected that the British would pursue, be baffled by their uncertainty as to the destination of the French, and that the latter would reach Europe again far enough in advance to control the Channel for some days. The plan failed through various causes. The British commander-in-chief before Toulon, Nelson, did follow the Toulon fleet to the West Indies; but, though starting a month later than it, the better quality of the British fleet enabled him to get back first, contrary to the Emperor's calculations. Nor could Napoleon reckon upon the singular insight with which Nelson at Antigua divined that Villeneuve had returned to Europe; so that, having left the Straits of Gibraltar thirty-one days after the allies, he sailed upon his return thither only four days later than they, and got back a week before they entered Ferrol.

I may add that, besides the principal diversion by his fleets assembling in the West Indies, Napoleon's correspondence at that time is full of schemes for enticing the British squadrons away from the Bay of Biscay and the Channel.

It will be observed that the French fleet, going in this instance to the West Indies, was intended to produce and did produce just that effect, which its going to Corfu would have produced upon conditions in the Mediterranean, in the case discussed in the last lecture. The British fleet was drawn in pursuit to the West Indies; that is, far from the Straits of Dover, the strategic center of Napoleon's plans, the critical point of the campaign, as Corfu was remote from Alexandria and from the line of communications between that port and France. If Nelson had been an average commander he would have remained in the West Indies until he had tangible evidence that the French fleet had left them. This is no surmise. Many strongly urged him so to remain; the weight of opinion was against him; but he possessed that indefinable sagacity which reaches just conclusions through a balancing of reasoning without demonstrable proof. If he had remained until he got reliable information, the result would have been twenty allies in Europe to support Napoleon's concentration, and the British concentration weakened by Nelson's dozen,—a total difference of thirty ships-of-the-line.

I will here draw your attention to the fact that Napoleon's

plan in this case was very similar to, and apparently derived from, one elaborated in 1762 by the French prime minister of that period.[2] Napoleon probably knew of this from the French archives; but it is unlikely that Nelson did.

Whichever of the above-mentioned two courses the navy has to adopt,—driving away or drawing away,—it must again be noted that it is on the defensive as to the general operations, but on the offensive as to its own actions. This, it may be further noted, is exactly what Napoleon prescribed to Marmont. "The turn which the general affairs of Europe have taken," he writes, "compels the Emperor to renounce for this year the expedition against Portugal," that is, an offensive campaign. Therefore he prescribes a general defensive attitude, but one carrying offensive menace; in order thereby to protect Badajoz, and to secure the line of communications from France to Madrid, which Salamanca covered, but which Sir John Moore three years before had threatened with such disastrous effect to Napoleon's own plans, drawing him away from his strategic center, the critical point of the campaign, at a moment vital to his success.

Now, such conditions of things, as regards a conquest actual or supposed, precisely illustrates also the relation of the navy to home defense. In both cases the nation, being in actual possession, is in so far on the defensive; but if, from necessity or by a mistaken policy, it keeps its navy also on the defensive within its ports or tied to them, it abandons to the enemy its commerce and its communications with abroad. This was what the United States perforce did in 1812, having no navy to send abroad, except as commerce destroyers. Such an abandonment will not necessarily lead to ruin, especially if the country be large, so as to be able to depend on internal resources, or have on its land frontiers neutral nations through whom a roundabout trade may be carried on. Nei-

2. Corbett, *Seven Years' War*, vol. ii, pp. 302–307.

ther will it altogether lose commerce in neutral bottoms, if its coast be too long for effective blockade; but it will suffer both humiliation and material loss, which a great nation should not risk. The true complement of any scheme of home coast defense is a navy strong enough either to drive the hostile fleet away from one's own shores or to keep it away by adequate threats to hostile interests. So used, a navy is unquestionably the best of coast defenses.

In this connection, because here entirely pertinent, I wish to introduce a comment which I shall develop at length in a later lecture; namely, that seacoast fortresses should not be thought, as they usually are, to be primarily defensive in function. Seacoast works, the office of which is limited to keeping hostile ships at a distance, but which are open on the land side, may be defensive merely; but a properly fortified port, capable of giving security to a navy, is defensive only as is a fortress, like Metz or Mayence, which contains an army able to take the field, and thereby compels the enemy to maintain before the place a detachment sufficiently strong to arrest any offensive action possible to the garrison.

Even our feeble War of 1812 yielded an illustration of this offensive character of a port capable of sheltering a squadron. The squadron of Commodore John Rodgers, in New York, was a garrison capable of acting offensively; and it did so. The British knew it had sailed in a compact body, but with what intentions they did not know. Consequently, their squadron on the American coast being small, they had to keep their ships together, lest one alone should encounter that squadron. This enforced concentration, coupled with the necessity for the British to protect their own trade, left the American ports so loosely watched that the greater part of the returning American merchant vessels arrived safely. This defensive result was intended, and was obtained, by Rodgers' offensive cruise, the credit of which, in conception and in execution, belongs almost wholly to him. The sustained exertion of such

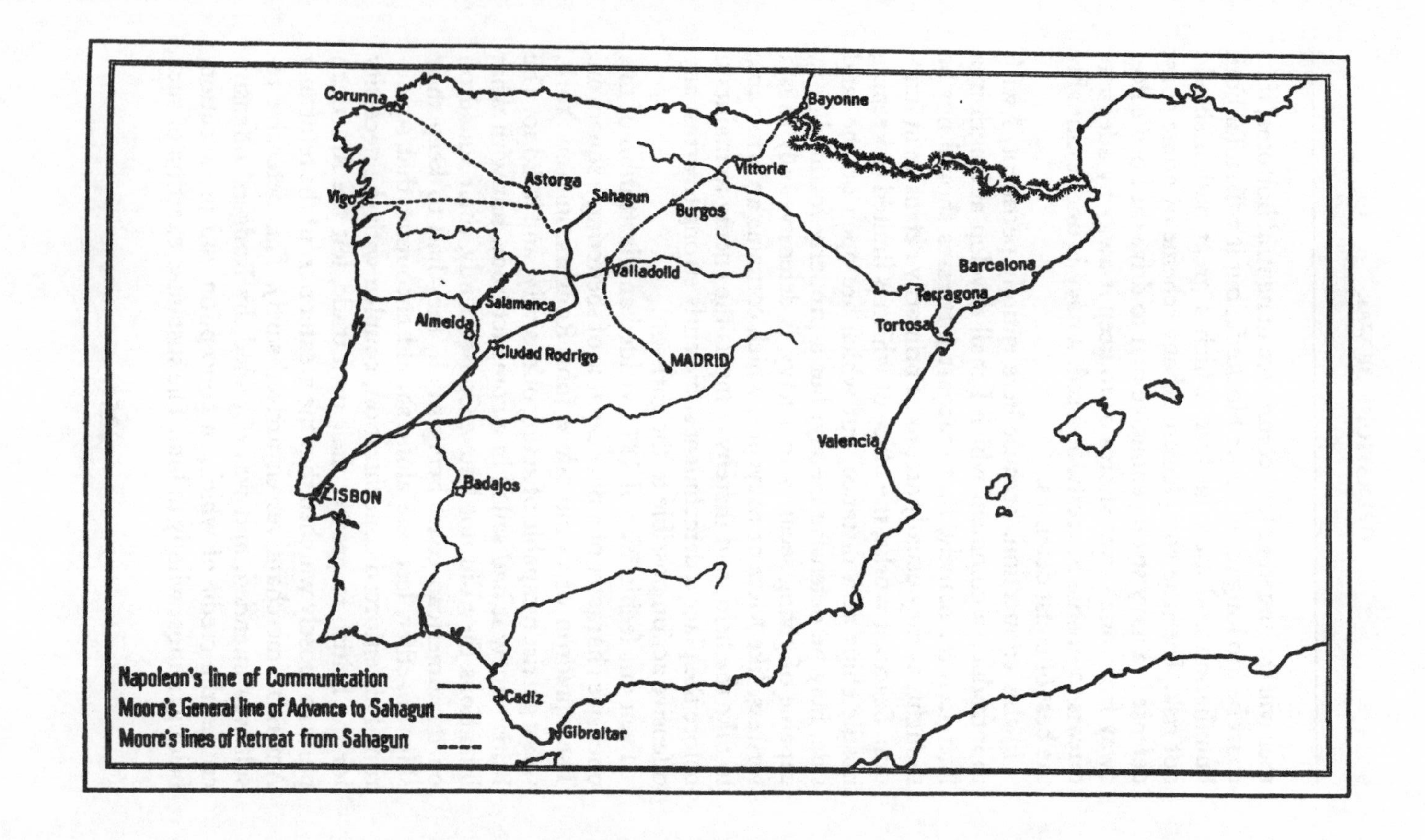
Corunna
Bayonne
Vittoria
Astorga
Sahagun
Vigo
Burgos
Valladolid
Barcelona
Tarragona
Salamanca
Almeida
Tortosa
Ciudad Rodrigo
MADRID
Valencia
Badajos
LISBON
Cadiz
Gibraltar
Napoleon's line of Communication
Moore's General line of Advance to Sahagun
Moore's lines of Retreat from Sahagun

offensive action depends upon ports capable of protecting the fleet. Otherwise it is destroyed, as at Port Arthur; or driven out, as at Santiago.

In this direction also we may seek a proper comprehension of what the size of a navy should be. It should be so great, and its facilities for mobilization and for maintenance of supplies should be such, that a foreign country contemplating war should feel instant anxiety because of the immediate danger that would arise from that navy, either to itself, or to its dependencies, or to its commerce. Such effect would be deterrent of war; and to deter is simply to practice diversion in another form. This has been announced, with military brevity and emphasis, as the official purpose of the German government in its naval programme adopted in 1900. "Germany must have a fleet of such strength that, even for the mightiest naval Power, a war with Germany would involve such risks as to jeopardize its own supremacy." Unhappily this purpose, when effected as towards Great Britain, will leave the United States far in the rear of Germany as a naval power.

It has been assumed, as a principle of strategy in reference to any theater of war, that the controlling point or system of points—the key of the situation, to repeat the familiar phrase—should be the objective of any offensive movement. It has also been stated in terms that the advance, or front of operations, should be pushed forward as far as can be done consistently with that closely linked communication, between all the parts, which binds the system into a whole. The reason is, that all within such a system, or in rear of such a front of operations, being in your control, is more useful to you than to the enemy; more dangerous to him than to you. This increases your resources for the time being; and if peace finds you in such possession, you have a vantage ground in that subsequent bargaining which is usually and euphemistically styled "negotiations for peace."

The supreme instance of such an advanced front of operations afforded by maritime warfare was the British blockade of the French and Spanish ports during the wars of the French Revolution and Empire, 1793–1815. The British fleets before the several ports,—Brest, Rochefort, Ferrol, Cadiz, Toulon,—linked together by intermediate divisions composed chiefly of cruisers, watching the smaller outlets and scouring the adjacent sea, formed really a continuous line or front of operations. The efficacy of such control was evidenced by the security of the British Islands and colonies, and of British trade upon the sea. The whole ocean, the region in rear of this front of operations, was secured against all but raiding. This was evidenced by the smallness of the loss by British commerce, less than three per cent of the total embarked; and by the failure of all the enemy's projects of invasion.

This advanced front held by fleets at sea corresponds to a front held by armies in the field, when maintained by virtue of their own superiority. It is evident, however, that such advantages will be increased by the holding of nearby fortified places. For example, when Bonaparte in 1796, in his advance against Austria, found himself held up by the fortress of Mantua, flanking his necessary line of advance, he took as his front of operations the line of the Adige, with the fortress of Verona bestriding the river. Verona was capable of withstanding a siege; it could be defended by relatively few men; it secured stores therein accumulated; it had also the moral prestige of strength, and it assured the passage of the stream when necessary to throw troops from one side to the other. All this was additional to the force of the French army itself. Thus at the battle of Arcola Bonaparte ventured to leave Verona with a very small garrison, while with the mass of his army, by night, he crossed the Adige lower down and struck at the rear of the Austrian army advancing on Verona. There was in this much risk and much bluff; but he was successful, which he could only have been by the use of the fortress. This served his army in the same manner that a fortified seaport serves a navy which dares strike outside in offensive-defensive action. By occupying this position, and by the additional strength conferred by the river and by Verona, prolonged to the north by Lake Garda, close along which the upper Adige flows, Bonaparte controlled all the resources of the valley of the Po and of south Italy, which lay behind; just as Great Britain did those of the ocean, by her fleets taking up the line of the French coast.

This maritime line likewise was strengthened by strong places; namely, by the home ports, Portsmouth and Plymouth, and abroad by Gibraltar, Malta, Port Mahon, and others. These were less open to attack than Verona was; and they afforded the local support of stores, of repairing, of refitting. Also, in case of sudden irruption by an escaping enemy's division they supplied refuge. Single ships or inferior divisions could find security within them. In addition to this defensive usefulness, such positions have also offensive power because of their nearness to, and thereby flanking, great lines of communication. Thus Gibraltar and Malta flanked all lines through the Mediterranean, Plymouth and Portsmouth through the Channel, Jamaica through the Caribbean. Similarly Brest, Cadiz, and so on flanked British lines to the southward; and therefore, as well as for other reasons, had to be contained, as Bonaparte had to contain Mantua.

Turning now to the Caribbean Sea in its entirety, as a region in which the United States might have occasion to desire influence and to exert it, and supposing all the islands at the outset to be in an enemy's possession, if Cuba should pass into our hands we should control a very important and useful position; but we might still be far from controlling thereby the whole sea. Assuming the opposing naval forces at the beginning to approach equality, it probably would not yet be in our power to control the whole. In that case the front of operations should be pushed as far forward as possible. For instance, it might be desirable to occupy Samana Bay and to control the Mona Passage; or, if strong enough, we might wish to push our front, the line maintained by the battle-fleet, to the southward and eastward, to harass the enemy and to protect the steamer routes through the Windward Passage to the Isthmus, which routes, by this advance of our own front, would lie in rear of our fleet.

Cuba, being now by the supposition ours, would cover our rear towards the Gulf of Mexico, wherein is an important part of the home base of operations. The base of the enemy may be, let us suppose, in the Lesser Antilles, and the sea between, under the supposition of equal navies, would be in dispute or in uncertain possession. The position of the enemy's fleet and its bases would indicate the direction of the next line of operations.

The enemy, being deprived of Cuba, whether by fortune of war or by original non-possession, might still hold Jamaica, as well as certain ports of the Windward Islands. This is the actual case of Great Britain, which holds Santa Lucia and Jamaica. In such case the main interest of the war would concentrate for the time around Cuba and Jamaica, which would become the critical point of the campaign. The series of posts, Great Britain, Santa Lucia, Jamaica, the Isthmus of Panama, would reproduce almost exactly the other existing line,—Great Britain, Gibraltar, Malta, the Isthmus of Suez. If the fleet at Jamaica should be inferior to that at Cuba, the Cuban fleet by taking position before Jamaica would intercept communications with the Windward Islands and reinforcements from them, would cover its own communications with Cuba and the United States, as well as the steamer routes, and by all this action would press the fleet within to battle, to relieve itself of these disadvantages. In this case as in the other the position of the enemy's fleet and his naval base indicate the direction of operations; as Port Arthur determined the direction of the Japanese naval war, as well as much of the Japanese effort on land, and Santiago that of the American fleet and army in 1898. The consequent movements of the Japanese and the Americans were a direct compulsion upon the Russians in the one instance and the Spaniards in the other to fight, which each avoided only by accepting fleet suicide.

Operations therefore should not cease with the occupation of the key. They should be pushed on untiringly; but the same reasoning which, to assure the hold on the key, prescribed the pursuit and destruction of the enemy's fleet holds good for the further operations. It is perhaps even more true of the sea than of the land that the proper objective is not a geographical point, but the organized military force of the enemy. Positions like Egypt and the defile of the Danube are important, not only nor mainly as inert masses of matter favorably placed, but because from them masses of trained warriors or of armed ships can act with such facility in different directions that they are worth more than greater numbers less well situated. The same is true of any place artificially fortified; its chief value is the facilitating the movements of the mobile forces. Therefore, with the possession of the advantages which the occupation of such positions confers comes the obligation to use them. How? The answer to this question is given in no doubtful manner by military writers. The organized force of the enemy, that is, his active army in the field, was the favorite objective of Napoleon, says Jomini.

Let it be supposed that you have seized such a strategic position and driven the enemy's fleet, after more or less fighting, off the theater of war in your rear and immediate front. This means that your home communications are secure, except from raiding, and that you have established your naval superiority for the time being. If the enemy's ships in an organized body, not as scattered cruisers, remain within the limits of the given theater in front of your present position, it will be because they still have points of support and supply, upon which they can depend for subsistence and to the defense of which they are necessary. It is not supposable that otherwise they can keep the sea within a restricted region; for the operations of coaling and taking on board stores, although probably feasible at sea if unmolested, could not go on with a superior fleet in the neighborhood, kept informed of their movements by watchful lookout ships and wireless telegraphy. Such points of supply or bases, therefore, there must be, and they indicate the direction of your next line of operations.

The usual great predominance of the British navy during the maritime historical period most vivid in our recollections has prevented naval strategy from yielding as many illustrations in point as otherwise might have been the case. The control which this predominance perpetuated over the communications between the enemy's bases and any objective proposed by him dried up at the source all other exhibitions of strategy, because communications, in the full meaning of the term, dominate war. As an element of strategy they devour all other elements. This usual predominance of a single Power gave origin to a French phrase of somewhat doubtful accuracy: *La Mer ne comporte qu'une seule maîtresse.* "The Sea brooks only one mistress." This is superficially plausible; but if understood to mean that the control of the sea is never in dispute, the mastery never seriously contested, it is very misleading. The control of the sea, even in general, and still more in particular restricted districts, has at times and for long periods remained in doubt; the balance inclining now to this side, now to that. Contending navies have ranged its waters in mutual defiance. This was conspicuously the case in the War of American Independence; to some extent also in the Seven Years' War, 1756–1763.

For instance, the attack upon Quebec and consequent reduction of all Canada in 1759 demanded as a first step the capture in 1758 of a fortress,—Louisburg, in Cape Breton. By this, the French fleet, which previously had gone back and forth, between France and Canada, in mass or in big detachments, was deprived of a necessary base of operations affecting the communications of the St. Lawrence River.

During the period of the French Revolution and of Napoleon, Great Britain was for most of the time sole mistress of the seas; yet in 1796 she was compelled to evacuate the Mediterranean. That limited area of the seas—the Mediterranean—was the scene of a protracted naval campaign with varying balance from 1793 to 1798, when it was decided finally in favor of Great Britain by the Battle of the Nile.[3] It is worth while to trace the leading incidents, for they illustrate the occurrence and the necessity of just such steps as we have been considering; thus proving that the consideration is not academic merely, but springs from the nature of things.

In 1793, the British fleet entered the Mediterranean, with the Spaniards as allies. Owing to the disloyalty of southern France to the Revolutionary government, an opportunity arose of obtaining possession of Toulon and its fleet; and this determined the first objective and line of operations of the allies. You will recall the importance attached to the capture of Toulon by the great Duke of Marlborough nearly ninety years before, and the strenuous though abortive attempt of Prince Eugène in 1707 to accomplish this result.

[Mahan here refers the reader to an earlier portion of *Naval Strategy*, pp. 97–99 of the original edition, where he had written:]

> . . . the allied navy [the Dutch and English], by pressure on the coast from Barcelona to Genoa, should support Savoy in closing the road by the valley of the Po to the French, to whom it offered a route alternative to the Danube valley for advance against Austria. Closing to France meant also keeping open for an Austrian army to move against Toulon. The reduction of this place was the real decisive object in the Mediterranean. It would give the allies a formidable port, a strategic position permanent for the existing war, immediately on

3. See map on pages 240 and 241.

their scene of naval operations, at the same time that the loss of it would paralyze the French navy locally; and it would remain a bridgehead for landings in southern France, the dread of which could not but detain a disproportionate number of French troops from reinforcing resistance to the allied armies in the Netherlands, or on the German frontier.

This was the broad underlying purpose of the naval campaign in 1704, in which Gibraltar fell. It failed, for reasons too complicated to detail here; but the influence of the fleet's presence upon Savoy, the pressure upon this flank of the French, contributed to favorable changes in the main theater of the war in Germany and on the upper Rhine. The maintenance of Savoy in her opposition to France depended upon support by the allied fleet, aided by troops of the coalition. This permitted Prince Eugene, who had commanded in Italy, to make in 1704, the year after Savoy's defection from France, the junction in the valley of the Danube with Marlborough, who had marched his army from the Netherlands south for this concentration; the result being the celebrated Battle of Blenheim, which inflicted upon the French a tremendous overthrow. This victory in turn relieved for the time the pressure upon Savoy, to accomplish which was one of Marlborough's objects in undertaking his great flank march; an interesting instance of the interaction of events in war.

Marlborough and Eugene persisted in their purpose against Toulon, which culminated in a direct attempt in 1707. This again failed; but the effect of this conjoint movement of the fleets and armies of the coalition, this flank attack, had been to cause so large a concentration of French troops in that quarter as to reduce France to inaction elsewhere. After this year the French abandoned Italy. Marlborough in 1708, after the mishap of Toulon, expressed his regret that the British ministry found it difficult to keep the fleet in the Mediterranean during winter. "Until it does so stay I am much persuaded you will not succeed in Spain." The want of a base other than Gibraltar was met by the capture of Minorca in the same year. It, with Gibraltar, was ceded to England at the peace. Minorca thus was a more useful conquest than Toulon, as Gibraltar was more than worth Cadiz; just because it was possible to obtain a cession, a permanent acquisition, which could scarcely have been done with either of the continental ports.

Thus, by obtaining for England fixed naval bases, the Peace of Utrecht made the strategic position of the Mediterranean permanently tenable by the British navy, conferring the power of acting upon the coast line everywhere, with the unforeseen and unforeseeable promptness which the mobility of naval force gives. From the particular territorial distribution of France and Spain, which entails on them commercial and military necessities on both the Atlantic and the Mediterranean, a superior navy, like that of England, by operating in the Mediterranean and towards its entrance, acts as did the Archduke on the Danube. The sea itself becomes a link, a bridge, a highway, a central position, to the navy able to occupy it in adequate force. It confers interior lines, central position, and communications militarily assured; but to hold it requires the possession of established bases, fortresses, such as those of which we have been speaking. Similarly, all these advantages followed the command of the American lakes, themselves Mediterraneans, in 1812–1814. "Without naval control of those lakes," wrote the Duke of Wellington, "successful land operations are impossible on that frontier."

The occupation of Toulon would paralyze at its source all French naval movement in the Mediterranean, and upon naval movement depended in great measure the land campaigns in northern Italy and along the Riviera. In 1793, Toulon was delivered by treachery; the allied fleets entered the port and allied troops occupied the lines surrounding it. The British admiral wished at once to seize or destroy the French squadron within, thus striking at the enemy's organized force; but this was opposed by his Spanish colleague, an officer of former wars, profoundly conscious and jealous of British naval superiority, which this destruction would increase. The British officer dared not chance the result of a rupture of the alliance; and this political consideration saved the French ships, most of which afterwards took part in the Battle of the Nile. Without them Bonaparte's Egyptian expedition could not have been started.

The French Government soon besieged Toulon, and the clear sight of Bonaparte into a tactical situation led to the seizure of a position from which batteries commanded the fleets at anchor. Toulon was evacuated perforce. The Spaniards retired to their ports; and the British, by the loss of the place itself, were compelled now to take the step which usually comes first. They had to obtain an advanced position, for refit and repair, for accumulation of stores,—in a word, a local base,—from which to control Toulon and support Austrian operations on the Riviera. This they had been able to neglect hitherto only through the chance which had placed Toulon itself in their hands. For a brief moment they took position in Hyères Bay, close to Toulon; but this was too near the mainland, too open to a repetition of the occurrences which had driven them from Toulon. Consequently they moved their advanced base to San Fiorenzo Bay, a harbor in the north of Corsica. This, with other ports of the island, they were enabled to occupy through the momentary disaffection of the inhabitants to France.

This advanced position they held, and garrisoned to an extent which was sufficient so long as the islanders sided with them. The waters between Corsica and Toulon and Genoa became a debatable ground, in which the British upon the whole predominated, but which could not be said to be in their undisputed control. The neighboring sea had perhaps only one mistress, but that mistress was not without a rival. The case is that of our supposed fleets, resting, one upon Cuba, the other on Santa Lucia or Martinique. Two fleet actions were fought with the Toulon ships in 1795; neither decisive. From San Fiorenzo as a base, operations were maintained along the Riviera, the ultimate objective of operations, in support of the Austrian advance against France; but here also nothing conclusive was effected. From 1794 to and including the first half of 1796 there was a perpetual conflict; resting on one side upon Toulon, on the other upon Gibraltar and upon the advanced base which the British had seized at San Fiorenzo. Concerning this period, Nelson some years after affirmed that, if the British admiral had been efficient, the French could not have maintained the forward position which they did. This, if accurate, means that when Bonaparte in April, 1796, took command of the army of Italy, he would have found the Austrians so far advanced, and the British navy in such control of the shore line from Nice to Genoa, that his plan of campaign must have been different. His very first movement, by which he struck in between and separated once for all the Austrians from their Piedmontese allies, was possible, only because the British and Austrian neglect of opportunities allowed him at the beginning to be at Savona, far in advance of Nice. Also, but for this, his communications, as well for troops—reinforcements—as for supplies and for ammunition, would have depended upon a very difficult land transport, with wretched roads, instead of the facile water route, following a coast-line bristling with French batteries.

You observe here the ultimate objective, northern Italy and the Riviera, the occupation of which by the allies of Great Britain would menace Toulon; you recognize also the intermediate objective, San Fiorenzo, essential for the maintenance of the British naval operations. Those operations thus constituted Nelson declared would have been successful under competent leadership. Ample time, two years, was afforded. Then came Bonaparte. With the advantages left in his hands, and by his own masterly management, in two months he had routed the Austrians and was on the Adige and in Verona. West and south of this position all opposition to him fell to pieces. The whole shore line of north Italy became French; and through its ports numerous French partisans passed in small boats to Corsica, stiffening there the disaffection to the British, which had begun already. At the same moment Spain, swayed largely by Bonaparte's victories, passed into alliance with France. A junction followed of the fleets of the two countries. Their organized naval forces, of which the detachment spared at Toulon was an important element, were combined. In face of this odds the British felt compelled to abandon their advanced position, and withdrew their fleet to Gibraltar, whence it further fell back upon Lisbon.

This movement of the organized force of the British navy was not molested by the allied fleets, which also separated soon after; the Spaniards going to Cartagena, the French to Toulon. The Spaniards then attempted to take their ships to Cadiz. The British admiral put to sea and met them off Cape St. Vincent. Though very inferior in numbers, he knew the superiority of his ships in quality. "A victory is very necessary for England," he was heard to say; and, the Spaniards offering him a tactical "opportunity," he fought and beat them. They retired into Cadiz, where the British shut them up by a blockade, the force of which was largely increased from home, in order to permit a detachment from it against the expedition rumored to be fitting out in Toulon. This detachment was entrusted to Nelson, who found the thirteen ships of the French off the coast of Egypt, where he annihilated them at the Battle of the Nile, in 1798. The organized Mediterranean naval force of France and Spain, having failed to utilize its opportunity when mistress of the northern Mediterranean, and having separated into two bodies, was thus beaten in detail, and the whole Mediterranean passed into British control for the rest of the war. Bonaparte having already compelled Austria to peace, in 1797, there were no longer any Austrian operations that the British fleet could assist; but its own front, dictated by the necessities of the time, ran from Gibraltar to Minorca, which had been seized by the commander-in-chief at Gibraltar when he received the news of the Nile, thence to

Sicily and Naples, and was continued on by way of Malta, still held by the French, to Alexandria and the Levant, where Bonaparte then was.

From this brief outline it is evident that the sea in the past has not been so exclusively dominated, even by Great Britain, at her greatest, that a contest for control may not take the form of a succession of campaigns marked by ups and downs. In fact, the very year after the Battle of the Nile a French fleet of twenty-five of the line broke into the Mediterranean, turned all the British dispositions upside down, and finally went back with fifteen or twenty Spaniards, bringing to Brest a concentrated body of over forty ships. Had there been any single port in the Mediterranean capable of maintaining such a body, they might have remained there; modifying conditions, if not reversing them. As it was, even in Brest they starved for want of supplies; but while this result shows the need of properly equipped bases, it also demonstrates that Great Britain did not control the sea in such sense that her hold could not be shaken. Yet in no war has she been more powerful at sea. In the War of American Independence, the West Indies and North America witnessed a like contest for control; of which Suffren's campaign in the East Indies, at the same period, is also a conspicuous illustration.

The necessity for properly equipped and properly situated local bases for a naval force in distant or advanced operations is also evident from this Mediterranean example. The War of American Independence offers a strikingly similar instance, on a smaller scale. In 1782 and 1783, the French and British squadrons in the East Indies were substantially equal. The scene of operations, dictated by local conditions on shore, was the Coromandel Coast; the east side of Hindustan. Naval coöperation was not practicable during the strength of the northeast monsoon, say from November to March, because the onshore wind blew with violence. The British retired for that season to Bombay, where they had establishments for re-

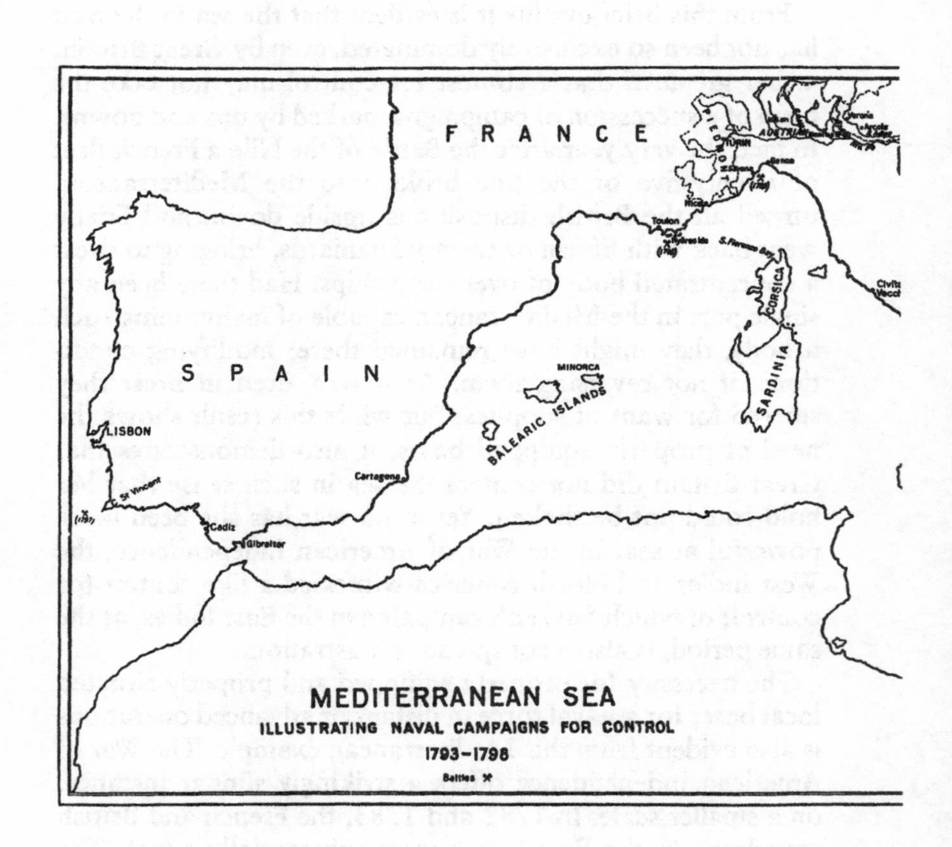

MEDITERRANEAN SEA
ILLUSTRATING NAVAL CAMPAIGNS FOR CONTROL
1793–1798
Battles ✕

TURKEY
Naples
CORFU
ASIA MINOR
Messina
MALTA
CRETE
CYPRUS
Aboukir Bay
ALEXANDRIA

fit. The French had no similar port nearer than the Mauritius, and all their previous campaigns had failed for want of an advanced base on, or near, the Coromandel Coast. It happened, on this occasion, that the Dutch, who then owned Ceylon, had entered the war against Great Britain in 1781, and the British had taken from them Trincomalee on the northeast coast of Ceylon; that is, in prolongation, southward, of the Coromandel Coast. There had not, however, been time enough to fortify the port properly, or else, as is more likely, the opportunity to do so had been neglected; and the French Admiral Suffren by dexterity and rapidity of action gained possession. This was the advanced position he needed, and when winter came he retired there to refit, thus remaining close to the scene of war, a very important matter for its political effect upon the French allies on shore. The British fleet went to Bombay, practically out of reach for four months or more. Had it retained Trincomalee, the French must have gone to Mauritius, or remained at anchor off a dangerous lee shore, with which they could seldom communicate. The result was that Suffren next season appeared on the scene two months before the British, and obtained successes which might have been decisive if peace had not intervened.

From the instances cited, it may be accepted that war upon the sea may take the shape of a protracted series of campaigns, between forces so nearly equal as to afford large play for strategic combinations. In fact, during the War of American Independence, of which Suffren's campaign was an episode, the seas of North America and the West Indies offered a similar illustration of the balance of naval forces; the scales swaying now to this side, now to that, up to the decisive events of Yorktown in 1781, and Rodney's victory of 1782, which marked the end of the struggle in either quarter. The same general result may be found, though to a less marked degree, even when one fleet is markedly superior to the other.

It is clear also that the vigor and celerity which are essential to ultimate success may depend upon the tenure of local bases of operations. This was one great advantage of Japan over Russia in the recent war. It may be, however, that the local bases of the enemy are too strongly fortified for attack, or that the assailant's force is too weak for quick reduction—as happened at Port Arthur.

As a matter of experience, local bases, if fairly fortified, are rarely attacked until the one party or the other has established predominance on the water. In 1760–1762, the British captured Martinique and Guadeloupe from the French, and again in 1794 and 1810; but in the intermediate war, of American Independence, they made no attempt upon either. This was partly because their army was occupied on the American continent; partly because they could not risk a big detachment of troops when a naval check, as at Grenada, in 1779, might cause its surrender. They did snatch away Santa Lucia; but that was at the very opening of hostilities, in 1778, and was due to their own local base at Barbados being reinforced before the French had completed their preparations for defense. It was a successful *coup de main*. In the same way, the French and Spaniards did not attempt Jamaica until 1782, when they were in superior force in the Caribbean. The attempt failed with Rodney's victory over the French fleet; that is, the allies were beaten in detail. The defeat of one part of their force rendered the other, the Spaniards, innocuous.

The distance from your own most advanced position to the position you wish to attack may be a further element of difficulty. To act from Cuba against the Windward Islands, such as Martinique and Santa Lucia, which by position control the eastern gateways of the Caribbean Sea, is clearly a more complicated undertaking than similar action from Cuba against Jamaica. If a harbor of Porto Rico is susceptible of adequate defense against attack in force, it would be in respect of situation a more advantageous base of operations against the

Windward Islands than anything in Cuba. St. Thomas, I believe, is capable of such defense; and it is better situated than Porto Rico.

To a fleet already weakened numerically by its first advance and by the necessity of guarding its first line of communications, such as that from the United States ports to Cuba, a second long line, as from Cuba to the Windward Islands, or to the Isthmus, is a serious consideration if the enemy be active. The care of the second advanced line may bring the fleet down to an equality of strength with the enemy who has an advantage in acting nearer his own base. In such a case the inability of the fleet to carry supplies, beyond a certain amount, in its own bottoms, should be supplemented by a depot—if any such offer—some distance in front of the position to which the first advance has been carried. Samana Bay or Porto Rico will illustrate such intermediate—that is, advanced—depots; corresponding to San Fiorenzo Bay as before cited, or to Port Royal and Key West in the War of Secession, or to Guantanamo in the American operations against Santiago. Such a depot need only be safe from a raid, for it may be assumed that the movements of the enemy can be watched sufficiently to prevent a sudden attack in force upon it; the fleet being, by the supposed advance, in face of the enemy's fleet and base. If more than one such advanced, or intermediate, point be available, a careful choice must be made between them; looking not only to their intrinsic advantages, but to their relations to the probable movements of the fleet, and to the first and second lines of communication, the protection of which will draw upon those forces of the fleet that ought to be dispersed as little as possible.

Having reference to the defense of the Canal, which is the crucial strategic feature of the whole Caribbean, it may be remarked that to attack an enemy's base, such as Martinique or Santa Lucia, is a more effective measure for protection and control of the Isthmus than a direct defense of the Isthmus itself would be; whether such defense be passive only, by fortification, or active, by a fleet resting upon the fortified Canal. For, if one of the islands—supposed an enemy's base—is attacked by a combined expedition, such attack, so long as sustained in adequate force, detains the scene of war at a distance from the Canal, and protects all communications west of the operation. It constitutes an advanced front of operations, combined with the moral power of the initiative and of the offensive. If it ultimately fails, it nevertheless will have produced this result for the time it lasts; while if successful, the enemy is deprived of a necessary base, the recovery of which involves operations that will exert the same protective influence as those which effected the capture.

If between the position you have first occupied and the enemy's base there is only clear sea (as between Cuba and Martinique, if not able to use Samana), it may be possible to take with the fleet a number of transports, above all, of colliers; especially if you are so superior as to allow a certain proportion of the fleet to be continually coaling, without reducing the number engaged below the enemy's strength. Before Trafalgar, Nelson thus sent his ships by groups of a half-dozen to water at Tetuan. Owing to the absence of such a division he had but twenty-seven instead of thirty-three on the day of battle. If he could have watered at sea, thus keeping his vessels together, the battle would have been even more decisive. The battleship Massachusetts lost her share in the battle of Santiago, because coaling at Guantanamo.

It will be borne in mind that a body of transports is always a tactical weakness in the day of battle, and will probably lower the fleet-speed of a number of high-powered ships of war. The question of speed in such an advance, however, may be of secondary importance, if the enemy's expected reinforcements cannot reach him within a known time; and, when about to engage, the safety of colliers and other incumbrances should be dismissed from mind, in view of the greater tactical necessity of beating the enemy while he is still inferior. In these calculations, much depends upon the respective strength of the two fleets. The best defense for the transports will be to attack the enemy and occupy him fully, just as the best defense of the Isthmus would be to attack an enemy's base. This has been, historically, the usual practice. In the expedition against Jamaica in 1782, De Grasse, when he found Rodney on his heels, sent his transports into the neighboring ports of Guadeloupe, and then fought.

The case of further advance from your new base may not be complicated by the consideration of great distance. The next step requisite to be taken may be short, as from Cuba to Jamaica; or it may be that the enemy's fleet is still at sea, in which case it is the great objective, now as always. Its being at sea may be because retreating, from the position you have occupied, towards his remoter base; either because conscious of inferiority, or, perhaps, after a defeat more or less decisive. It will then be necessary to act with rapidity, in order to cut off the enemy from his port of destination. If there is reason to believe that you can overtake and pass him with superior force, every effort to do so must be made. The direction of his retreat is known or must be ascertained, and it will be borne in mind that the base to which he is retreating and his fleet are separated parts of one force, the union of which must be prevented. In such a case, the excuses frequently made for a sluggish pursuit ashore, such as fatigue of troops, heavy roads, etc., do not apply. Crippled battleships must be dropped, or ordered to follow with the colliers. Such a pursuit presumes but one disadvantage to the chasing fleet, viz., that it is leaving its coal base while the chase is approaching his; and this, if the calculations are close, may give the pursuing admiral great anxiety. Such anxieties are the test and penalty of greatness. In such cases, excuses for failure attributed to shortness of coal will be closely scrutinized; and justly. In all other respects, superiority must be assumed, because on no other condition could such headlong pursuit be made. It aims at a great success, and successes will usually be in proportion to superiority, either original or acquired. "What the country needs," said Nelson, "is the annihilation of the enemy. Only numbers can annihilate."

If such a chase follow a battle, it can scarcely fail that the weaker party—the retreating party—is also distressed by crippled ships, which he may be forced to abandon—or fight. Strenuous, unrelaxing pursuit is therefore as imperative after a battle as is courage during it. Great political results often flow from correct military action; a fact which no military commander is at liberty to ignore. He may very well not know of those results; it is enough to know that they may happen, and nothing can excuse his losing a point which by exertion he might have scored. Napoleon, says Jomini, never forgave the general who in 1796, by resting his troops a couple of hours, failed to get between an Austrian division and Mantua, in which it was seeking refuge, and by his neglect found it. The failure of Admiral de Tourville to pursue vigorously the defeated Dutch and English fleet, after the battle of Beachy Head, in 1690, caused that victory to be indecisive, and helped to fasten the crown of England on the head of a Dutch King, who was the soul of the alliance against France. Slackness in following up victory had thus a decisive influence upon the results of the whole war, both on the continent and the sea. I may add, it has proved injurious to the art of naval strategy, by the seeming confirmation it has given to the theory of the "fleet in being." It was not the beaten and crippled English and Dutch "fleet in being" that prevented an invasion of England. It was the weakness or inertness of Tourville, or the unreadiness of the French transports.

Similarly, the refusal of Admiral Hotham to pursue vigorously a beaten French fleet in 1795, unquestionably not only made that year's campaign indecisive, but made possible Napoleon's Italian campaign of 1796, from which flowed his whole career and its effects upon history. The same dazzling career received its sudden mortal stab when, in the height of his crushing advance in Spain, with its capital in his hands, at the very moment when his vast plans seemed on the eve of accomplishment, a more enterprising British leader, Sir John Moore, moved his petty army to Sahagun, on the flank of Napoleon's communications between France and Madrid. The blow recoiled upon Moore, who was swept as by a whirlwind to Coruña, and into the sea; but Spain was saved. The Emperor could not retrieve the lost time and opportunity. He could not return to Madrid in person, but had to entrust to several subordinates the task which only his own supreme genius could successfully supervise. From the military standpoint, his downfall dates from that day. The whole career of Wellington, to Waterloo, lay in the womb of Moore's daring conception. But for that, wrote Napier, the Peninsular War would not have required a chronicler.

An admiral may not be able to foresee such remote consequences of his action, but he can safely adopt the principle expressed by Nelson, in the instance just cited, after hearing his commander-in-chief say they had done well enough: "If ten ships out of eleven were taken, I would never call it well enough, if we were able to get at the eleventh."

The relations between the fleets of Admirals Rozhestvensky and Togo prior to their meeting off Tsushima bore no slight resemblance to those between a pursued and a pursuing fleet. The Russian fleet, which had started before the Port Arthur division succumbed, was placed by that event in the position of a fleet which has suffered defeat so severe that its first effort must be to escape into its own ports. This was so obvious that many felt a retreat upon the Baltic was the only course left open; but, failing that, Rozhestvensky argued that he should rush on to Vladivostok at once, before the Japanese should get again into the best condition to intercept him, by repairing their ships, cleaning the bottoms, and refreshing the ships' companies. Instead of so ordering, the Russian government decided to hold him at Nossi-Bé (the north end of Madagascar), pending a reinforcement to be sent under Admiral Nebogatoff. Something is to be said for both views, in the abstract; but considering that the reinforcement was heterogeneous and inferior in character, that the Russian first aim was not battle but escape to Vladivostok, and, especially, that the Japanese were particularly anxious to obtain the use of delay for the very purpose Rozhestvensky feared, it seems probable that he was right. In any event, he was delayed at Nossi-Bé from January 9 to March 16; and afterwards at Kamranh Bay in French Cochin China, from April 14 to May 9, when Nebogatoff joined. Allowing time for coaling and refitting, this indicates a delay of sixty to seventy days; the actual time underway from Nossi-Bé to Tsushima being only forty-five days. Thus, but for the wait for Nebogatoff, the Russian division would have reached Tsushima two months before it did, or about March 20.

Togo did not have to get ahead of a flying fleet, for by the fortune of position he was already ahead of it; but he did have to select the best position for intercepting it, as well as to decide upon his general course of action: whether, for instance, he should advance to meet it; whether he should attempt embarrassment by his superior force of torpedo vessels, so as to cripple or destroy some of its units, thus reducing further a force already inferior; also the direction and activities of his available scouts. His action may be taken as expressing his opinions on these subjects. He did not advance; he did not attempt harassment prior to meeting; he concentrated his entire battle force on the line by which he expected the enemy must advance; and he was so far in ignorance of their movements that he received information only on the very morning of the battle. This was well enough; but it is scarcely unreasonable to say it might have been bettered. The Japanese, however, had behind them a large part of a successful naval campaign, the chief points of which it is relevant to our subject to note. They had first by a surprise attack inflicted a marked injury on the enemy's fleet, which obtained for them a time of delay and opportunity during its enforced inactivity. They had then reduced one of the enemy's two naval bases, and destroyed the division sheltered in it. By this they had begun to beat the enemy in detail, and had left the approaching reinforcement only one possible port of arrival.

If a flying fleet has been lost to sight and has but one port of refuge, pursuit, of course, will be directed upon that port; but if there are more, the chasing admiral will have to decide upon what point to direct his fleet, and will send out despatch vessels in different directions to find the enemy and transmit intelligence. Cruisers engaged in such duty should be notified of the intended or possible movements of the fleet, and when practicable should be sent in couples; for although wireless telegraphy has now superseded the necessity of sending one back with information, while the other remains in touch with the enemy, accidents may happen, and in so important a matter it seems expedient to double precautions. The case resembles duplicating important correspondence; for wireless cannot act before it has news, and to obtain news objects must be seen. It is to be remembered, too, that wireless messages may be intercepted, to the serious disadvantage of the sender. It seems possible that conjunctures may arise when it will be safer to send a vessel with tidings rather than commit them to air waves.

Thus, in theory, and to make execution perfect,—to capture, so to say, Nelson's eleventh ship,—the aim must be to drive the enemy out of every foothold in the whole theater of war, and particularly to destroy or shut up his fleet. Having accomplished the great feature of the task by getting hold of the most decisive position, further effort must be directed towards, possibly not upon, those points which may serve him still for bases. In so doing, your fleet must not be divided, unless overwhelmingly strong, and must not extend its lines of communication beyond the power of protecting them, unless it be for a dash of limited duration.

If compelled to choose between fortified ports of the enemy and his fleet, the latter will be regarded as the true objective; but a blockade of the ports, or an attack upon them, may be the surest means of bringing the ships within reach. Thus, in the War of American Independence, the siege of Gibraltar compelled the British fleet on more than one occasion to come within fighting reach of the enemy's blockading fleet, in order to throw in supplies. That the allies did not attack, except on one occasion, does not invalidate the lesson. Corbett in his Seven Years' War points out very justly, in Byng's celebrated failure, which cost him his life, that if he had moved against the French transports, in a neighboring bay, the French admiral would have had to attack, and the result might have been more favorable to the British. Such movements are essentially blows at the communications of the enemy, and if aimed without unduly risking your own will be in thorough accord with the most assured principles of strategy. A militarily effective blockade of a base essential to the enemy will force his fleet either to fight or to abandon the theater of war. Thus, as has been pointed out elsewhere, in Suffren's campaign in Indian Seas, so long as Trincomalee was in possession of the British, a threat at it was sure to bring them out to fight, although it was not their principal base. The abandonment of the theater of war by the navy will cause the arsenal to fall in time, through failure of resources, as Gibraltar must have fallen if the British fleet had not returned and supplied it at intervals. Such a result, however, is less complete than a victory over the enemy's navy, which would lead to the same end, and so be a double success, ships *and* port.

If the enemy have on the theater of war two or more ports of supplies, which together form his base, and those points fulfill the condition before laid down, that they should not be so near that both can be watched by one fleet, the task becomes more difficult. The two most important naval stations on our Atlantic coast, Norfolk and New York, offer such conditions, being some two hundred and fifty miles apart; and to a retreating United States fleet the second entrance to New York, by Long Island Sound, together with Narragansett Bay, constitutes for a pursuing enemy a further complication which favors escape. A single port with widely separated entrances approaches the condition of two ports, in the embarrassment imposed upon an enemy who has lost touch. Admiral Togo was confronted with just this perplexity. Vladivostok could be reached by three different routes, wide apart. A position heading off all three could be found close before Vladivostok itself; but, besides the possibility that an unfavorable chance, such as a fog, might allow the Russians to slip by, in which case they would not have far to go to get in, there was also the risk that, even if defeated, those which escaped for the moment could enter, thus making a victory less decisive. The pursuit of the day following the battle picked up ships which had got by, and in the supposed case might have reached port.

In the instructions of Napoleon to Marmont, before cited, the Emperor, estimating the various chances from the dispositions he has ordered, considers that of battle near Salamanca. This is to be desired, he writes, because, if beaten so far from the sea, the English will be ruined, and Portugal thereby conquered. This distance from the sea was distance from the English refuge. It was the merit of Sir John Moore that in the headlong pursuit by the Emperor he avoided decisive action, and got his army to the sea; fugitive, and demoralized by exhaustion, but still saved. It is to be remembered that in this very recent instance of Tsushima the Japanese immediately before the battle had lost touch with the enemy. Over a century before Togo, Rodney failed more than once to intercept French bound to Martinique, because they used one of the many passages open between the Windward Islands to enter the Caribbean, and so escaped detection until too late to be intercepted. To cruise before Martinique was ineffective, because the French had other available refuges in Guadeloupe; and besides, with the perpetual trade wind, and calms intervening, blockading sailing ships fell to leeward,—could not keep their station.

The guiding principle in all these cases is that your force must not be divided, unless large enough to be nowhere inferior to the enemy, and that your aim should be to reduce his base to a single point, out of which he can then be driven by regular operations, or by exhaustion; or, at least, to reach which with supplies, or for refuge, his fleet must accept battle. Thus the British in 1794, and again in 1808–1810, took from the French both Martinique and Guadeloupe, depriving them of all foothold in the West Indies, and so securing the Caribbean for British commerce. As regular operations usually require much more time than assault does, if there be more than one arsenal the weaker would preferably be carried by force, leaving the strongest to fall afterwards by the less dangerous means of regular operations already indicated for a single post. In the Mediterranean, from 1798 to 1800, the French held both Malta and Egypt. The strength of the fortifications of Malta is known; whereas Egypt had nothing comparable. Egypt after long blockade was reduced by an extensive combined assault, a great fleet and a considerable army. Malta was overcome by cutting its communications, and surrendered to exhaustion. Port Arthur was taken by force. If Rozhestvensky had reached Vladivostok without a battle, the war would have continued; but the Japanese under the conditions would probably have contented themselves with blockading that port, relying upon the presence of their fleet assuring all the sea behind, which would secure the communications of the army in Manchuria.

Going on thus from the simpler to the more difficult cases,

we now reach the one where your strength is not great enough to give present hope of driving the enemy quite out of the field of war. That is, an attitude generally defensive in character succeeds one that is distinctively offensive. When this occurs, you will seek to occupy as advanced a position as possible, consistently with your communications. Such advanced position may not be a point or a line of the land, but at sea. If Cuba, for instance, belonged to the United States, it may be conceived that the fleet would seek to control the Mona Passage, resting on no nearer base than the eastern ports of Cuba. Or from the same base, the fleet might seek to maintain a cruising ground to the southward in the Caribbean Sea, to harass the enemy's commerce or protect the interests of its own nation. It might, again, in its advanced position, be simply waiting for an expected attack,—an attempt, possibly, of the enemy to regain the position which had been seized,—when its duties would be to delay, harass and finally attack him, as before suggested in speaking of his action upon your advance.

The object of such an advanced position as is now being discussed is to cover the ground or sea in its rear, and to meet and hinder the advance of the enemy. Consequently, it should be chosen in all its details with strict reference to strategic considerations; until circumstances change, a farther permanent advance is not contemplated. The position taken up, therefore, has reference to the lines of communication behind it and to those by which it is approached from the side of the enemy; those which it covers and those by which it is threatened. Napoleon, in 1796, made the line of the Adige his front of defense, thereby covering all the ground in his rear and securing the use of it for the supply of his army. Thus, too, the British fleet at San Fiorenzo Bay, Corsica, in 1794–1796 rested on that bay as a base, and maintained thence its front of operations before the gates of Toulon. This kept in check a powerful French fleet in the port, covered the communications with Gibraltar, and secured the Mediterranean for British trade.

When too much is not risked thereby, the line should be advanced to include cross-roads or narrow passages that are near to. Though the open sea has not natural strategic points, yet the crossing of the best mercantile routes, the difficulties of strong head winds and adverse currents, will make some points and some lines more important than others. The occurrence of strong harbors, possibly shoal water, or other difficulties to navigation, may affect the tracing of the line laid down to be held. A fleet, for instance, advanced to the Mona Passage and resting upon no nearer fortified port than those in Cuba, might yet venture to establish in Samana Bay a depot of coal, which would facilitate its remaining on the ground, yet the loss of which would not be a vital injury, in case of defeat. An advance of the enemy being expected, everything that delays or that makes farther advance hazardous is useful. The fleet, it cannot be too often repeated, is the chief element of strength in naval warfare; but the fleet with strong points to support it is stronger than the fleet alone.

We have now brought our expeditionary fleet, which has hitherto been on the offensive and advancing, to a standstill. The efforts which it has made, the losses which it has undergone, by battle, or by detachments necessitated by its lengthening lines, the difficulties in front, all or some of these lay upon it the necessity of stopping for a time, as did Napoleon in the case I have just cited. This stoppage will be for the purpose of securing conquests made; of strengthening the supply ports in the new base, so that the defense of them may be thrown upon the land forces, thus releasing detachments of ships hitherto tied to them; of storing in these ports supplies in such quantities as to be independent, for a long time, of the mother country and of the first line of communications which connects with it. When Bonaparte had established himself at Verona and on the Adige, he not only had useful control of all

Italy south and west of that position, except the besieged fortress of Mantua; he also had placed the communications nearest to France so remote from interruption, that detachments once necessary to guard them were no longer required for that purpose. The communications were as if in France itself.

These processes amount to a military occupation of the conquered positions, incorporating the conquest militarily with the home country; and will result in releasing the navy, in great measure, from the direct defense of the conquered ports, in which at first it will have to aid. By such establishment of the advanced position, dependence upon the original lines of communications is lessened, and the burden of defending them diminished. The detachments thus released will join the fleet, and, with other reinforcements sent from home, may so increase its strength as to enable it again to take up the direct offensive; a step which will be made upon the same general strategic principles as have already been given for the first advance.

It will remain, therefore, to consider more particularly the principles governing a defense, which have at times been alluded to in speaking of the opponent's action during your own advance.

Suggestions for defense cannot be as satisfactory, superficially at least, as those for offense, because the defensive is simply making the best of a bad bargain; doing not the thing it would like to do, but the most that can be done under the circumstances.

It is true that in certain respects the defensive has advantages, the possession of which may even justify an expression, which has been stated as a maxim of war, that "Defense is a stronger form of war than Offense is." I do not like the expression, for it seems to me misleading as to the determinative characteristics of a defensive attitude; but it may pass, if properly qualified. What is meant by it is that in a particular operation, or even in a general plan, the party on the defense, since he makes no forward movement for the time, can strengthen his preparations, make deliberate and permanent dispositions; while the party on the offensive, being in continual movement, is more liable to mistake, of which the defense may take advantage, and in any case has to accept as part of his problem the disadvantage, to him, of the accumulated preparations that the defense has been making while he has been marching. The extreme example of preparation is a fortified permanent post; but similar instances are found in a battle field carefully chosen for advantages of ground, where attack is awaited, and in a line of ships, which by the solidarity of its order, and deployment of broadside, awaits an enemy who has to approach in column with disadvantage as to train of guns. In so far, the *form* taken by the defense is stronger than the *form* assumed for the moment by the offense.

If you will think clearly, you will recognize that at Tsushima the Japanese were on the defensive, for their object was to stop, to thwart, the Russian attempt. Essentially, whatever the tactical method they adopted, they were to spread their broadsides across the road to Vladivostok, and await. The Russians were on the offensive, little as we are accustomed so to regard them; they had to get through to Vladivostok—if they could. They had to hold their course to the place, and to break through the Japanese,—if they could. In short, they were on the offensive, and the form of their approach had to be in column, bows on,—a weaker form,—which they had to abandon, tactically, as soon as they came under fire.

In our hostilities with Spain, also, Cervera's movement before reaching Santiago was offensive in character, the attitude of the United States defensive; that is, he was trying to effect something which the American Navy was set to prevent. There being three principal Spanish ports, Havana, Cienfuegos, and Santiago, we could not be certain for which he would try, and should have been before two in such force that an attempt by him would have assured a battle. We were strong enough for such a disposition. The two ports thus to be barred were evidently Havana and Cienfuegos. The supposed necessity for defending our northern coast left Cienfuegos open. Had Cervera made for it, he would have reached it before the Flying Squadron did. The need for keeping the Flying Squadron in Hampton Roads was imaginary, but it none the less illustrates the effect of inadequate coast defenses upon the military plan of the nation.

The author whom I quote (Corbett, *Seven Years' War,* vol. i, p. 92), who himself quotes from one of the first of authorities, Clausewitz, has therefore immediately to qualify his maxim, thus:

> When we say that defense is a stronger form of war, *that is, that it requires a smaller force, if soundly designed,* we are speaking, of course, only of one certain line of operations. If we do not know the general line of operation on which the enemy intends to attack, and so cannot mass our force upon it, then defense is weak, because we are compelled to distribute our force so as to be strong enough to stop the enemy on any line of operations he may adopt.

Manifestly, however, a force capable of being strong enough on several lines of operation to stop an enemy possesses a superiority that should take the offensive. In the instance just cited, of Cervera's approach, the American true policy of concentration would have had to yield to distribution, between Cienfuegos and Havana. Instead of a decisive superiority on one position, there would have been a bare equality upon two. Granting an enemy of equal skill and training, the result might have been one way or the other; and the only compensation would have been that the enemy would have been so badly handled that, to use Nelson's phrase, he would give no more trouble that season, and the other American division would have controlled the seas, as

Togo did after August 10, 1904. From the purely professional point of view it is greatly to be regretted that the Spaniards and Russians showed such poor professional aptitude.

The radical disadvantage of the defensive is evident. It not only is the enforced attitude of a weaker party, but it labors under the further onerous uncertainty where the offensive may strike, when there is more than one line of operation open to him, as there usually is. This tends to entail dissemination of force. The advantages of the defensive have been sufficiently indicated; they are essentially those of deliberate preparation, shown in precautions of various kinds. In assuming the defensive you take for granted the impossibility of your own permanent advance and the ability of the enemy to present himself before your front in superior numbers; unless you can harass him on the way and cause loss enough to diminish the inequality. Unless such disparity exists, you should be on the offensive. On the other hand, in the defensive it has to be taken for granted that you have on your side a respectable though inferior battle fleet, and a sea frontier possessing a certain number of ports which cannot be reduced without regular operations, in which the armed shipping can be got ready for battle, and to which, as to a base, they can retire for refit. Without these two elements there can be no serious defense.

The question for immediate consideration here, however, is not the defense of the home coast, but the defense of a maritime region of which control has been acquired, wholly or in part. Unless this region be very close to the mother country, the power of the nation will not be as fully developed and established as at home. The nearness of the Caribbean Sea gives special value to any judiciously placed acquirements of the United States there,—such as the Panama Canal Zone, Porto Rico, and Guantanamo, as compared with the same in possession of European states. So also the position of Japan in the Farther East confers on her a very marked advantage over every European or American state for sustaining and compacting her power, and for carrying on operations of war. But where the intervening distances are very nearly equal, the maritime region of our present hypothesis lies between the two distant contestants as a debatable land, very much as Germany and the Danube Valley lay in former days between Austria and France. This was the case with the region embracing the West Indies and the thirteen American colonies, during the maritime war associated with that of American Independence. The islands and the continent, with the intervening seas, were the principal scene of the maritime war, and they were substantially equidistant from the great Powers engaged: France, Great Britain, and Spain. On land, the control of such a remote region depends upon two elements: the holding of certain points as bases, and the maintenance of an active army in the field; but it is according as the army is stronger or weaker than that of the enemy that it takes the offensive or defensive. Now, in a maritime region, the navy is the army in the field.

It is in the defensive that the strong places play their most important part. When an army is advancing in superior force, those belonging to it will be behind—in its rear. They serve then as safe points for the assembling of supplies, of trains, of reinforcements. If well garrisoned, and in secure communication with the army, the latter maneuvers freely.

> "It is desirable," says the Archduke Charles, speaking more particularly of the base of operations—but the remark applies also to intermediate points,—"that these points should be fortified so as to be able to leave them to themselves without fear of losing the magazines there established, and not to be obliged to defend them with detachments, which have always the inconvenience of weakening the army. The movements of the general-in-chief, forced before all to cover his magazines and to leave troops to guard them, will never be as

> rapid nor as bold as if he had the faculty of moving away from them for some time with the certainty of finding them again intact."

The same is true of any naval base of operations, if inadequately defended; and the more useful and necessary it is to the fleet, the greater the hindrance to naval movements which may expose it, when so undefended.

The Archduke spoke from sad experience; if not his own personally, at least that of the armies of his country. The Austrians, probably because they were of Germanic blood, akin to the countries in which they were operating, did not live off them in the unscrupulous manner then practised by the French. They needed, therefore, large depots, for which they could not always have a fortified town. Consequently, they either had to leave a large body of men to guard them, which weakened the main army, or, if they tried to cover them with the latter, its freedom of movement was seriously impaired.

Note that those extreme advocates of the navy as a coast defense, who decry fortifications, would put the navy in a similar predicament.

It will not be inferred from this that the most strongly fortified places do not demand garrisons; but the strength of the walls represents so many men, and, moreover, troops of a quality unfit for the field may man works. It is just so with a seaport; if it has no fortifications, the navy may have to undertake a great part of the defense; if adequately fortified, the detachment from the navy is released and the defense carried on by troops not fitted for service afloat. Such places are the foundation upon which an offensive best rests; and in regions permanently belonging to a nation, they should be so chosen, with reference to intrinsic fitness and relative position, as to be coördinated into a strategic system, to the power of which each contributes. They must not be too few; neither must they be too many, for to protect and garrison them takes from the numbers of the active army—the army in the field. When, therefore, the number of fortresses exceeds that which is necessary, the active army is not strengthened, but weakened. "France," says Jomini, "had too many fortified places, Germany too few; and the latter were generally bad" (that is, weak) "and unsuitably placed." Under these conditions, it is not to be wondered at that the experience of the French officer and the German Archduke led the former rather to depreciate and the latter to exaggerate the value of fortified posts.

This question of fortified points of support, depots for supply, and if need be for momentary refuge, assumes peculiar importance in the matter immediately before us—the control of a maritime region external to the country. In such a case the army in the field is preëminently the navy. The land forces will commonly be confined to holding these positions, defensively; expeditionary, or offensive, movements will be for them exceptional. A fleet charged with the protection of such bases, whether at home or abroad, is so far clogged in its movement, and is to the same extent in a false position. An egregious instance at the present moment is the fear in Great Britain of German invasion. This is due to the great inferiority of the army in the British Islands to that of Germany. The British Islands are inadequately garrisoned; they depend for defense upon the fleet alone; and the fleet consequently is tied to British waters. If Great Britain on her own soil could meet Germany man to man, equal in numbers and in training, the fleet would have relatively a free foot. It could afford, for example, to spare a detachment to the Mediterranean, or to China; retaining at home only a reasonable superiority to a possible enemy. As things are, since all depends upon the fleet, the fleet must have a wider margin of safety, a crushing superiority; that is, its freedom of movement and range of action are impaired greatly, by the necessity of keeping with it ships which under other conditions might be spared.

A navy may be thrown perilously upon the defensive in its

general action in a particular region, because obliged to cover two or more points inadequately protected by fortification or army. Thus, in 1799, the unexpected entrance of twenty-five French battleships into the Mediterranean turned everything upside down, because so many points were thought to need protection, and could receive it only from the fleet, because they were inadequately garrisoned; a precise reproduction, on a smaller scale, of the present dilemma of the British Islands. Specifically, the British commander-in-chief felt the weight of Minorca; and he used concerning it an expression which is worthy of remembrance as bearing upon the Fleet in Being theory and the Blue Water School. "It is too bad," he said, "that I cannot find these vagabonds,"—that is, the French fleet,—"and that I am so shackled by the care of this *defenseless* island." The man who used these words was not a commander of the first order; but he was an officer of more than usual distinction, of capacity proved much above the average, and he here expressed a frame of mind inevitable to the average man. But for the necessity of protecting positions, the British fleet would have concentrated, and would have moved freely at will, and in force, offensively against the enemy. As it was, not knowing the enemy's purpose, it was kept in two principal divisions, neither of them equal to the French whole. One, the main body, covered Minorca and kept moving somewhat aimlessly in the triangle defined by Barcelona, Toulon, and Minorca; the other, under Nelson, covered the approaches to Naples and Sicily. Amid this maze of British perplexity arising from the need of defending several points with the fleet alone, the French acted safely, though hastily, and retired unmolested; taking with them from Cartagena a large body of Spanish ships, which remained in Brest hostages for the alliance of Spain.

It is when thrown on the defensive, that the value of strong places is most felt. The first object, in order, of the defense, is to gain time. It is therefore of advantage that opposition to the enemy should begin as far as possible in front of the vital points of the defense. In Bonaparte's famous campaign in Italy, of 1796, the dexterity of his strategy and the audacity of his tactics enabled him in two months to cover the ground from Savona, well west of Genoa, to Mantua. There the fortress held him for nine months. Observe that Lombardy, the valley of the Po, then an appendage of Austria, was to Austria just what an external maritime region, the Caribbean for instance, may be to a maritime state like the United States. Though long since lost, Austria perhaps has never reconciled herself to a surrender which shut her off from the Mediterranean, a new approach to which she seems now to be seeking in the Balkans. Mantua was to her an advanced post, which by its effect on the movements of an approaching invader protected, not only the region wherein it itself lay, but the home country behind. Its powerful garrison, like a fleet in a seaport, threatened, and unless checked would control, the French communications as they advanced up the Alps towards the Austrian home territory. Bonaparte had not force enough to oppose to the garrison, and at the same time to move forward. He could not divide, so had to stop; and in the nine months of delay Austria collected and sent against him no less than three successive armies, whom he repelled only by a display of skill, daring, and energy, of which he alone was capable. Two months after Mantua fell, he had progressed so far towards Vienna that Austria asked for peace.

A fortress like Mantua, in a case like this, affords a striking instance of defense being a strong form of war, and also of the advantage of opposition to an enemy's approach beginning as far as possible in advance of the home territory. It is perhaps an extreme example. Yet, all the while, Bonaparte was showing how much stronger in spirit, and in effect, offense is; for, while holding his position in Verona and on the Adige, which was his base of defense, it was by rapid offensive movements, resting on these positions, that he disconcerted the enemy, who, being taken continually by surprise by the French initiative, was forced on the defensive and ultimately compelled to retreat. To the offensive belongs the privilege as well as the risks of the initiative; and the distinguishing value of the initiative is that its purpose, known to itself, is one, is concentrated. The defense, being ignorant of its opponent's purpose, to which it is compelled to conform its dispositions, feels endangered in more directions than one. Its tendency therefore is to dissemination, as that of the offense is towards concentration.

It is to be observed that conditions like Mantua may not always exert the like effect upon the movements of a fleet, because ships carry in their holds much of that which communications mean to an army. For example, in 1801, after the battle of Copenhagen and destruction of the Danish fleet, Nelson wished to proceed at once up the Baltic against a strong Russian naval detachment lying at Revel; but his commander-in-chief was unwilling to advance, leaving Denmark still hostile in his rear and unsubdued. This could mean only sensitiveness about communications, which for such an enterprise was pedantic; because the going and coming would not have exhausted the British resources, whereas the destruction of the Russian division was of military and political importance. Nelson admitted the risk, but urged the superior necessity of taking it. He was overruled, and the Russians escaped. Such a rapid dash has something of the nature of raids, which characteristically disregard communications. If, instead of such sudden attack, the purpose had been a prolonged operation,—a blockade for instance, such as then maintained before French ports,—supplies for the fleet in the Baltic must pass within gunshot of Danish batteries, which therefore must be reduced. If, besides the batteries, a Danish naval division were there, a British naval division must balance it also. If the Revel and Copenhagen bodies, taken together, had equalled or exceeded the British numbers, division of these would have been inexpedient; and Copenhagen must first be subdued, as Bonaparte had to reduce Mantua. For modern fleets, the exigencies of coal renewal aggravate such a situation.

When inferior to the enemy, an army in the field must fall back, disputing the ground if possible, until the advanced line of fortified strategic points is reached. As it passes that line, it has to strengthen the points in proportion to their needs, to its own present strength, and to its hopes of reinforcement. To shut itself up in one of the fortresses, as Mack at Ulm, McMahon at Sedan, Bazaine at Metz, whether justified by the situation or not, is a counsel of despair, so far as that army is concerned. The general military situation may require the step, but it is a confession of disaster. When the pursuing enemy reaches the line of fortified posts, the question presents itself to him: "Shall this point be taken before going farther, or shall I leave only enough force here to prevent its garrison acting against my communications?"

If the decision be to besiege, time is lost; if to proceed, the pursuing army is weakened relatively to the pursued. This weakening process goes on with each place observed, but the pursuer may be better able to stand it than the pursued. An inferior force outside, not intending to besiege, may adequately check a superior distributed in two or more places, because the different detachments cannot combine their movements, and the inferior has the advantages of central position and interior lines. Moreover, the pursuer is necessarily superior, and may be greatly superior; and as he passes on he endangers or destroys the lines of supply to the place, the fall of which then becomes a question of time. These considerations show both the value and the limitations of fortified points. Their passive strength, however great, can never bring about the results which may be attained by a skillfully handled army in virtue of its mobility.

Warfare at sea does not seem to present a very close analogy to the case of an inferior army retreating before a superior, disputing its progress by resistances which take advantage of successive accidents of the ground, that contribute to the stronger "form" of war which defense is. Yet we have historical parallels, which to say the least are suggestive. Such is that of Nelson off Sicily in 1799, with less than a dozen battleships, expecting the approach of a French supposed nineteen,—actually twenty-five,—and intending to fight rather than let them occupy the places he was set to defend; and again, in 1805, when returning from the West Indies to Europe with twelve ships and expecting to meet eighteen to twenty enemies. In both cases he was animated with the same purpose, expressed in the words, "By the time they have beaten my division, they will give no more trouble this year." He meant, of course, that his share in the whole action of the British navy was one incident in the process of beating the enemy in detail; leaving the rest of the British force to finish the remainder. This corresponds essentially with the action of the southern Austrian force of the Archduke Charles, in 1796, which had as its share to occupy Moreau by retreat, fighting at every defensible point, while the Archduke himself with the northern troops turned upon Jourdan in overwhelming numbers.

Doubtless, too, in the first instance, Nelson had in his mind the same purpose that he explicitly stated in the second: "I will not fight until the very last moment, *unless* they give me an opportunity too tempting to be resisted,"—that is, a clear momentary advantage. An advantage is an advantage, however offered or obtained; whether by an enemy's mistake, or by the accidents of the ground that play so large a part in land war; and on either element a skillful defense looks warily for its opportunities to the enemy's mistakes, as well as to other conditions. Napoleon is reported to have said at Austerlitz, when urged to seize an evident opportunity, "Gentlemen, when the enemy is committed to a mistake, we must not interrupt him too soon." The comprehensive rôle of the British navy in the wars of Nelson's time was defensive, and in broad strategic lines it followed Bonaparte's practice; that is, its dispositions were such as to constitute original advantage, to assume the offensive promptly when occasion arose, and to fight at advantage when opportunity offered. So, when leaving the Mediterranean early in 1805 to pursue the French to the West Indies, Nelson met a convoy of reinforcements for Malta, a defensive measure. Pressed as he was for time, he waited till all arrangements for its safe arrival were perfected. He looked out for his bases of defense, while himself bound on an errand of offense.

At sea, as on land, fortified posts are necessary. Their importance is perhaps even greater, because the field in which fleets act rarely offers positions—due to the contour of the ground—by which an inferior force by tactical dispositions can lessen the odds against it. The need of refuge, and of security for resources, is greater. The old advantage of the wind is represented by greater speed, and the fleet speed of a few ships is likely to be greater than that of a larger number. The more numerous the ships of one fleet, the more likely to be found among them, not only the fastest, but also the slowest in the two forces; and fleet speed is not an average, but the speed of the slowest. The fleet speed of the more numerous fleet is consequently likely to be less. This consideration shows that precipitate flight to the support of its ports may not be necessary to the retreating fleet, especially if, as is possible, the approaching navy is convoying land forces in transports.

Still, considering the open nature of the field, it may be said that the retreating fleet, if greatly inferior, should not let the assailant get within striking distance. There seems reason to say that it should fall back, proportioning its speed to that of the approach, with fast cruisers in its rear keeping sight of the enemy, and with communication established between them and the main body. The enemy's light vessels will, of course, try to drive these off, but they cannot follow them into their own fleet, nor can they prevent their return. Granting equal speed, the cruisers of the pursuing fleet cannot overtake. They can only keep those of the retreating at a certain distance from the pursuers' main body, which will be of less advantage, because their own presence reveals the nearness of their main force. The retiring cruisers must not fight, unless with special advantage; because, if crippled, they will fall into the hands of the approaching enemy. The utmost, then, that we can say for the weaker fleet of the defendant, under these circumstances, is that it should keep as near the invader as feasible, waiting to seize any advantage that may turn up. *How* to seize such advantage belongs rather to the province of tactics; as, indeed, does the whole conduct of such a retreat. Granting equal speed and professional skill to begin with, a smaller number can generally move more rapidly than a larger, and are more easily handled. How the larger should move, in what order, how protect its convoy; how the smaller should conduct its retreat, what possibilities of harassing attack are presented by modern conditions, and the best method of making them—all these things belong to the province of Grand Tactics, rather than of Strategy.

When the retreating fleet has reached the outer line of its fortified ports,—the first line of defense,—the two parts of the defendant's force, his fleet and ports, are united. The question then arises of the use to be made of the fleet. The approaching enemy is, by the supposition, superior on the sea; and also on land, at least as to the particular objective he has first in view. If there be but a single port of the defendant's, the case is very serious, for his supply of coal becomes precarious. If the single port is so ill fortified that it cannot hold out a respectable time, the situation, as regards the particular region, is almost desperate.

As, however, there is no object in discussing desperate

cases, but only those where inferiority is not so great but that skill and activity may partially compensate for the inequality, we will assume that there are two or more ports reasonably defended, in position to afford support to each other, yet not so near together that an enemy can watch both without dividing his fleet. The aim now of the defendant's fleet—the weaker fleet—is threefold: the battleships should be kept together; they should endeavor not to be shut up in either port; and the battle fleet should not allow itself to be brought to action by the superior force, unless favored by circumstances. If uncertain as to the point first aimed at by the enemy, it will take the most favorable position for reaching either and wait further indications. Thus Nelson, at such a moment of uncertainty, as to whether the French fleet, escaped from Toulon, were bound to Egypt, or to the Atlantic, wrote, "I will neither go to the eastward of Sicily, nor to the westward of Sardinia, until I know something positive." Togo at Masampo affords another illustration; but less striking, because with fewer elements of doubt.

In choosing its local base of action, its point of concentration for the general defensive, of which it itself is a principal factor, the defendant fleet should consider seriously, among other things, which port is most likely to be the object of the enemy's shore operations; because, if that be ascertained, some other position will probably be better for itself. Thus, there were several reasons for presuming that the Japanese would prefer to land in the neighborhood of Port Arthur, and would attack that place. Consequently, the Russian fleet, if intending to postpone battle, or to decline it, would be better in Vladivostok; because, by taking position in Port Arthur it enabled, and even induced, the enemy to concentrate both fleet and army at one point, which thus became strategically, though not geometrically, a central position, occasioning the Japanese no temptation to eccentric movements. The Russian battle fleet at Vladivostok would draw thither necessarily the main Japanese fleet, and so would open larger possibilities to the Russian cruising divisions for action against the communications of the Japanese army. That Vladivostok has two entrances is an additional reason.

If the first objective be strong enough to require prolonged operations to reduce it, the enemy's fleet will be tied to that point wholly or in part. Even if it take no share in the direct attack, it will have to cover the communications of its army at their point of arrival, the most critical link of the chain connecting the army with home, and must block the use of the port to the defendant's shipping as a coaling or supply station. Only the fall of the place can wholly release the fleet of the assailant. It therefore will have two duties: one, to support the land attack, the other, to check any mischief set on foot by the defendant's navy. If the latter be wise and active, both duties cannot be attempted without some division of the attacking fleet. In the case supposed, the admiral of the defendant fleet enjoys the advantage of the initiative, in that his object is only one, however many ways of compassing it may offer. This advantage he has, because, although his country is on the defensive, and therefore his fleet also, the particular function of the fleet in the general scheme of defense is to take the offensive against the enemy's communications, or against his detachments, if such are made; in general, to divert and distract. As against these diversions and alarms, the fleet of the offense has to defend. Therefore it has two necessary objects, viz.: the hostile fleet and the hostile port, unless the defendant plays into his hands by letting his fleet be caught in the besieged port, as the Russians did at Port Arthur.

Let us suppose, for example, that the line of defense for a United States fleet was the Atlantic coast,—with the two ports of Norfolk and New York well fortified,—the United States navy inferior, but still strong. If the great commercial importance of New York determined the enemy's attack there, the United States fleet being in Norfolk would constitute two objects for the hostile navy and impose a division of force; otherwise, the United States navy, being left free to act, might attack any of the enemy's interests—trade, communications, colonies. If New York were the enemy's objective, and had but one entrance, it would, in my opinion, be a mistake to put the fleet there; but with two, they by themselves impose divergence. The introduction of wireless telegraphy will modify these considerations; but in view of weather conditions, and of the total advantages attendant upon the initiative, namely, that the choice of time, place, and manner is with the departing fleet, wireless can only modify, cannot annul.

It can scarcely be repeated too often that when a country is thrown on the defensive, as regards its shore line, the effectual function of the fleet is to take the offensive. Hence, in another part of this course, I have said that coast fortresses are not essentially defensive in character, as commonly esteemed, but offensive; because they guard the navy which is to act offensively. The instance of John Rodgers' squadron in 1812, though on so petty a scale, remains entirely in point. The United States, having almost no navy, nor army, was on the defensive; but the sailing of Rodgers' squadron was a step of general offense against British trade and naval detachments. Consequent upon it, British detachments had to concentrate, because weaker than Rodgers' whole; and American ports remained open for the returning merchant ships. Forgetfulness or neglect of this consideration was a leading factor in the Russian mismanagement of their fleet. It is immaterial whether the defensive is the original attitude of the nation, as in these cases of Russia and the United States, or whether it results from defeat upon the sea and retreat to home waters. When retreat is over, and the opportunity of harassing the advancing enemy, whether well or ill improved, has passed away, the defendant fleet is tied down to nothing except keeping the bunkers full; a weighty exception, it will be admitted, which we owe to steam. Still, it is a great thing to have no other cares, to be tied to no other duties.

On the supposition of proper fortification for the ports of the coast line, they are able to look out for themselves during a given time. The duty of the defendant admiral, then, is to strike at the communications of the enemy; to harass and perplex his counsels by attacks or threats, in every possible direction; to support the general defensive, by himself taking the offensive. The skill of the admiral or government charged with the direction of such operations will be shown by the choice of those objects of attack which will most powerfully move the enemy. The history of war is full of instances where sound military principles have been overridden by political or sentimental considerations, by lack of military skill in the commanders of fleets and armies, or of moral courage to bear a great responsibility. The object of the defense will be to play upon such weaknesses of human nature, with a view to make the offensive divide his forces. The impulse to try to protect every point can only be overcome, like other natural infirmities, by sound principles firmly held. At the time of our hostilities with Spain, the Navy Department was besieged with applications from numerous points of the coast for local protection. The detention of the Flying Squadron at Hampton Roads, as well as of a patrol force on the North Atlantic coast which would have been better employed in blockade and dispatch duty, may be considered concessions to this alarm. They certainly were not in accord with sound military principle.

The result aimed at by such operations of the defendant navy has been styled "displacement of force" by a recent French writer on Naval Strategy, Commander Daveluy. The phrase appears to me apt and suggestive. By it he means that, assuming the enemy to have disposed his forces on sound military principles, he is to be provoked, or allured, or harassed, or intimidated, into changing those dispositions, into displacing his forces. Over-confidence may be as harmful as over-caution, in inducing displacement. If properly concentrated, the hostile ships may be moved to disseminate; if correctly posted, to remove to a worse position. The capture of the Guerrière by the Constitution was due to a displacement of British forces. Rodgers' sailing in squadron had compelled the British to concentrate, and for the same reason to convoy an important West Indies fleet several hundred miles eastward in the Atlantic. There it was thought safe to detach the Guerrière to Halifax. On her way she met the Constitution.

I quote from Daveluy a few paragraphs:

> The maritime defensive, from whatever point of view regarded, offers only disadvantages. It may be imposed; it never should be voluntarily adopted. On the one side as on the other, we are led to choose the offensive; that is to say, to seek the enemy with the object of fighting him. But the two parties will not do this by the same methods.
>
> The stronger will hasten to meet the different divisions of the enemy, in order to destroy them before they have time to injure him. The weaker—whom I have called the defendant—will seek first of all to withdraw from touch of the enemy in order by uncertainty as to the points threatened to effect displacement of the hostile force, and to give rise to the unforeseen; then he will try to draw his enemy to a field of battle where his own feebler units can come into play advantageously. So long as this stage lasts, and until a decisive action has inclined the balance definitely, the immediate objects of the war are postponed to the necessity of first engaging the enemy under favorable conditions. In this game, the more active, the more skillful, the more tenacious and the better equipped will win.
>
> At the opening of a war especially, the offensive will produce decisive results. If successful in anticipating the projects of the enemy by impetuosity of attack, the general operations receive the predetermined direction; a situation is created

> which overthrows all the enemy's expectations, and paralyzes him, unless he succeed in retrieving his condition by a victory. The very fact of being forced into an unexpected situation puts him in a state of inferiority, and prevents him from recovery, while at the same time your own forces can be better utilized.

This effect was strikingly produced upon the Russians by the first successful surprise by the Japanese.

> The characteristic of the offensive is that it makes the attack instead of accepting it; this is evidenced in history by the fact that almost all naval victories have been gained upon the enemy's coast.

If, in the shock of war, all things in both sides were equally strong, there could be no result. On the other hand, when inequality exists, the weaker must go down before the stronger. It is in converting inequality or inferiority into superiority at a given point that the science, or rather the art, of war consists. The principles upon which this art is based, we are assured by the best authorities, are few and simple; and they are summed up in one great principle, that of being superior to the enemy at the decisive point, whatever the relative strength of the two parties on the whole. Thus the Russian navy in the aggregate was much superior to the Japanese, but, being divided, was inferior to the enemy upon the immediate scene of war; and this inferiority at the decisive point was increased by the sudden action of the Japanese in opening hostilities.

It is in the application of sound general principles to particular problems of war that difficulty arises. The principles are few, the cases very various, the smaller details almost infinitely numerous. Here experience enters—experience which, under the other form of the word, *experiments,* lies at the basis of all our science. But how shall experience of war be acquired in the absence of a state of war? And even amid constant war, how shall any one man, particularly a subaltern or naval captain, find in his own experience all, or any large portion of the innumerable cases that may and do arise? No one will answer that he can so find them; but if one be found bold enough to affirm he can, I throw myself back upon the words of great captains. The Archduke Charles writes:

> A man can become a great captain only with a passion for study and a long experience. There is not enough in what one has seen oneself; for what life of a man is fruitful enough in events to give a universal experience; and who is the man that can have the opportunity of first practicing the difficult art of the general before having filled that important office? It is, then, by increasing one's own knowledge with the information of others, by weighing the conclusions of one's predecessors, and by taking as a term of comparison the military exploits, and the events with great results, which the history of war gives us, that one can become skillful therein.

The first Napoleon similarly says:

> Make offensive war as did Alexander, Hannibal, Cæsar, Gustavus Adolphus, Turenne, Prince Eugene, Frederick the Great; read and reread the history of their eighty-three campaigns, model yourself upon them; it is the only means of becoming a great captain and of surprising the secrets of the art. Your mind thus enlightened will make you reject maxims opposed to those of these great men. . . . The history of these eighty-three campaigns, carefully told, would be a complete treatise on the art of war; the principles which should be followed in offensive and defensive war would flow from it as from a spring.

Again he says:

> Tactics, evolutions, the science of the engineer and of the artillerist, may be learned in treatises, almost like geometry; but knowledge of the great operations of war is acquired by experience, and by the study of the history of wars and of the battles of great captains.

There is yet another and deeper thought underlying the advice to study the campaigns of great commanders. It is not merely that the things they have done become a catalogue of precedents, to which a well stored memory can refer as special cases arise for decision. Such a mechanical employment of them has its advantage, can be consigned to treatises, and can be usefully taught to those who will learn nothing otherwise. But, beyond and above this, it is by that diligent study which Napoleon enjoins that the officer who so lives with those men absorbs not merely the dry practice, but the spirit and understanding which filled and guided them. There is such a thing as becoming imbued with the spirit of a great teacher, as well as acquainted with his maxims. There must indeed be in the pupil something akin to the nature of the master thus to catch the inspiration,—an aptitude to learn; but the aptitude, except in the rare cases of great original genius, must be brought into contact with the living fire that it may be itself kindled.

It is something like this, doubtless, that Napoleon meant when he drew the distinction quoted, between the elementary parts of the art—tactics, evolutions, etc.—and the conduct of great operations, which can be acquired only by experience and in the study of history. This he elsewhere expresses in a warning against dogmatizing on such matters:

> Such questions, propounded even to Turenne, to Villars, or to Prince Eugene, to Alexander, Hannibal, or Cæsar, would have embarrassed them greatly. To dogmatize upon that which you have not practised is the prerogative of ignorance; it is like thinking that you can solve, by an equation of the second degree, a problem of transcendental geometry which would have daunted Lagrange or Laplace.

Jomini, who fully agrees with the two leaders about the study of history, expresses the same idea by saying that the successful conduct of war is not a science, but an art. Science is sure of nothing until it is proved; but, all the same, it aims at absolute certainties,—dogmas,—towards which, through numerous experiments, it keeps moving. Its truths, once established, are fixed, rigid, unbending, and the relation between cause and effect are rather laws than principles; hard lines incapable of change, rather than living seeds. Science discovers and teaches truths which it has no power to change; Art, out of materials which it finds about it, creates new forms in endless variety. It is not bound down to a mechanical reproduction of similar effects, as is inanimate nature, but partakes of the freedom of the human mind in which it has its root. Art acknowledges principles and even rules; but these are not so much fetters, or bars, which compel its movements aright, as guides which warn when it is going wrong. In this living sense, the conduct of war is an art, having its spring in the mind of man, dealing with very various circumstances, admitting certain principles; but, beyond that, manifold in its manifestations, according to the genius of the artist and the temper of the materials with which he is dealing. To such an effort dogmatic prescription is unsuited; the best of rules, when applied to it, cannot be rigid, but must have that free play which distinguishes a principle from a mere rule.

Maxims of war, therefore, are not so much positive rules as they are the developments and applications of a few general principles. They resemble the every varying, yet essentially like, forms that spring from living seeds, rather than the rigid framework to which the free growth of a plant is sometimes forced to bend itself. But it does not therefore follow that there can be no such maxims, or that they have little certainty or little value. Jomini well says,

> When the application of a rule and the consequent maneuver have procured victory a hundred times for skillful generals, shall their occasional failure be a sufficient reason for entirely denying their value and for distrusting the effect of the study of the art? Shall a theory be pronounced absurd because it

has only three-fourths of the whole number of chances in its favor?

Not so; the maxim, rooting itself in a principle, formulates a rule generally correct under the conditions; but the teacher must admit that each case has its own features—like the endless variety of the one human face—which modify the application of the rule, and may even make it at times wholly inapplicable. It is for the skill of the artist in war rightly to apply the principles and rules in each case.

It is thus we must look upon all those rules of war that are advocated before us. The teacher who, without the tests of large experience, dares to dogmatize, lays himself open to the condemnation pronounced by Napoleon. But, on the other hand, men who deliberately postpone the formation of opinion until the day of action, who expect from a moment of inspiration the results commonly obtained only from study and reflection, who hope for victory in ignorance of the rules that have generally given victory, are guilty of a yet greater folly, for they disregard all the past experience of our race.

I end with an apposite quotation from the Archduke Charles:

> A general often does not know the circumstances upon which he has to decide, until the moment in which it is already necessary to proceed at once with the execution of the necessary measures. Then he is forced to judge, to decide, and to act, with such rapidity that it is indispensable to have the habit of embracing these three operations in a single glance, to penetrate the consequences of the different lines of action which offer, and to choose at the same moment the best mode of execution. But that piercing perception which takes in everything at a glance is given only to him who by deep study has sounded the nature of war, who has acquired perfect knowledge of the rules, and who has, so to speak, identified himself with the science. The faculty of deciding at once and with

certainty belongs only to him who, by his own experience, has tested the truth of the known maxims and possesses the manner of applying them; to him alone, in a word, who finds beforehand, in his positive acquirements, the conviction of the accuracy of his judgments. Great results can be obtained only by great efforts.

"Upon the field of battle," says the great Napoleon, "the happiest inspiration is most often only a recollection."

CHAPTER VIII

CONSIDERATIONS GOVERNING THE DISPOSITION OF NAVIES

WE have the highest military authority for saying that "War is a business of positions"; a definition which includes necessarily not only the selection of positions to be taken, with the reasonings, or necessities, which dictate the choice, but further also the assignment of proportionate force to the several points occupied. All this is embraced in the easy phrase, "The distribution of the fleet." In these words, therefore, ought to be involved, by necessary implication, an antecedent appreciation of the political, commercial, and military exigencies of the State in the event of possible wars; for the dispositions of peace should bear a close relation to the contingency of war. All three elements form a part of the subject-matter for consideration, for each is an essential factor in national life. Logically separable, in practice the political, commercial, and military needs are so intertwined that their mutual interaction constitutes one problem. The frequent statement that generals in the field have no account to take of political considerations, conveys, along with a partial truth, a most misleading inference. Applied even to military and naval leaders, it errs by lack of qualification; but for the statesman, under whom the soldier or seaman acts, the political as well

National Review, vol. 39 (July 1902), 701–19. Reprinted in *Retrospect & Prospect* (Boston, 1902), 139–205.

as the military conditions must influence, must at times control, and even reverse, decision.

The choice of situations, localities, to be held as bases of operations, is governed by considerations of geographical position, military strength, and natural resources, which endure from age to age; a permanence which justifies the expense of adequate fortification. The distribution of mobile force, military or naval, is subject to greater variation, owing to changes of circumstances. Nevertheless, at any one historical moment, of peace or war, this question also admits of an appropriate fixed determination, general in outline, but not therefore necessarily vague. This conclusion should be the outcome of weighing the possible dangers of the State, and all the various factors—political, commercial, and military—which affect national welfare. The disposition thence adopted should be the one which will best expedite the several readjustments and combinations that may be necessitated by the outbreak of various particular wars, which may happen with this or that possible enemy. Such modification of arrangements can be predicated with reasonable certainty for a measurable period in advance. The decision thus reached may be called the "strategic" solution, because dependent upon ascertainable factors, relatively permanent, of all which it takes account; and because also it is accepted, consciously and of purpose, as preliminary to the probable great movements of war, present or prospective.

In the particular cases that afterwards arise from time to time, and of which the outbreak of war may itself be one, the unforeseen, the unexpected, begins to come into operation. This is one of the inevitable accompaniments of warfare. The meeting of these new conditions, by suitable changes of plan, is temporary in character, varying possibly from day to day; but it will generally be found that the more comprehensive has been the previous strategic study, and the more its just forecasts have controlled the primary disposition,—the distribution of force,—the more certainly and readily will this lend itself to the shifting incidents of hostilities. These movements bear to the fundamental general dispositions the relations which tactics have to strategy. In them, on occasions, one or two of the leading considerations which have each had their full weight in the original dispositions, may have to be momentarily subordinated to the more pressing demand of a third. In war, generally and naturally, military exigencies have preponderant weight; but even in war the safety of a great convoy, or of a commercial strategic centre, may at a given instant be of more consequence than a particular military gain. So political conditions may right be allowed at times to overweigh military prudence, or to control military activity. This is eminently true, for, after all, war is political action. The old phrase, "The cannon is the last argument of kings," may now be paraphrased, "War is the last argument of diplomacy." Its purpose is to compass political results, where peaceful methods have failed; and while undoubtedly, as war, the game should be played in accordance with the well-established principles of the art, yet, as a means to an end, it must consent to momentary modifications, in accepting which a well-balanced mind admits that the means are less than the end, and must be subjected to it.

The question between military and political considerations is therefore one of proportion, varying from time to time as attendant circumstances change. As regards the commercial factor, never before in the history of the world has it been so inextricably commingled with politics. The interdependence of nations for the necessities and luxuries of life have been marvellously increased by the growth of population and the habits of comfort contracted by the peoples of Europe and America through a century of comparative peace, broken only by wars which, though gigantic in scale, have been too short in duration to affect seriously commercial relations. The unmolested course of commerce, reacting upon itself, has contributed also to its own rapid development, a result furthered by the prevalence of a purely economical conception of national greatness during the larger part of the century. This, with the vast increase in rapidity of communication, has multiplied and strengthened the bonds knitting the interests of nations to one another, till the whole now forms an articulated system, not only of prodigious size and activity, but of an excessive sensitiveness, unequalled in former ages. National nerves are exasperated by the delicacy of financial situations, and national resistance to hardship is sapped by generations that have known war only by the battlefield, not in the prolonged endurance of privation and straitness extending through years and reaching every class of the community. The preservation of commercial and financial interests constitutes now a political consideration of the first importance, making for peace and deterring from war; a fact well worthy of observation by those who would exempt maritime commercial intercourse from the operations of naval war, under the illusory plea of protecting private property at sea. Ships and cargoes in transit upon the sea are private property in only one point of view, and that the narrowest. Internationally considered, they are national wealth engaged in reproducing and multiplying itself, to the intensification of the national power, and that by the most effective process; for it relieves the nation from feeding upon itself, and makes the whole outer world contribute to its support. It is therefore a most proper object of attack; more humane, and more conducive to the objects of war, than the slaughter of men. A great check on war would be removed by assuring immunity to a nation's seaborne trade, the life-blood of its power, the assurer of its credit, the purveyor of its comfort.

This is the more necessary to observe, because, while commerce thus on the one hand deters from war, on the other hand it engenders conflict, fostering ambitions and strifes which tend towards armed collision. Thus it has continuously been from the beginning of sea power. A conspicuous instance was afforded by the Anglo-Dutch wars of the seventeenth century. There were other causes of dissatisfaction between the two nations, but commercial jealousies, rivalry for the opening markets of the newly discovered hemispheres, and for the carrying trade of the world, was the underlying national, as distinguished from the purely governmental motive, which inspired the fierce struggle. Blood was indeed shed, in profusion; but it was the suppression of maritime commerce that caused the grass to grow in the streets of Amsterdam, and brought the Dutch Republic to its knees. This too, it was, that sapped the vital force of Napoleon's Empire, despite the huge tributes exacted by him from the conquered states of Europe, external to his own dominions. The commerce of our day has brought up children, nourished populations, which now turn upon the mother, crying for bread. "The place is too strait for us; give place where we can sell more." The provision of markets for the production of an ever-increasing number of inhabitants is a leading political problem of the day, the solution of which is sought by methods commercial and methods political, so essentially combative, so offensive and defensive in character, that direct military action would be only a development of them, a direct consequent; not a breach of continuity in spirit, however it might be in form. As the interaction of commerce and finance shows a unity in the modern civilized world, so does the struggle for new markets, and for predominance in old, reveal the unsubdued diversity. Here every state is for itself; and in every great state the look for the desired object is outward, just as it was in the days when England and Holland fought over the Spice Islands and the other worlds newly opening before them. Beyond the seas, now as then, are to be found regions scantily populated where can be built up communities with wants to be supplied; while elsewhere are teeming populations who may be led or manipulated to recognize necessities of which they have before been ignorant, and stimulated to provide for them through a higher development of their resources, either by themselves, or, preferably, through the exploitation of foreigners.

We are yet but at the beginning of this marked movement, much as has been done in the way of partition and appropriation within the last twenty years. The regions—chiefly in Africa—which the Powers of Europe have divided by mutual consent, if not to mutual satisfaction, await the gradual process of utilization of their natural resources and consequent increase of inhabitants, the producers and consumers of a commerce yet to be in the distant future. The degree and rate of this development must depend upon the special aptitudes of the self-constituted owners, whose needs meantime are immediate. Their eyes therefore turn necessarily for the moment to quarters where the presence of a population already abundant provides at once, not only numerous buyers and sellers, but the raw material of labor, by which, under suitable direction and with foreign capital, the present production may be multiplied. It is not too much to say that, in order further to promote this commercial action, existing political tenure is being assailed; that the endeavor is to supplant it, as hindering the commercial, or possibly the purely military or political ambitions of the intruder. Commercial enterprise is never so secure, nor so untrammelled, as under its own flag; and when the present owner is obstructive by temperament, as China is, the impulse to overbear its political action by display of force tends to become ungovernable. At all events the fact is notorious; nor can it be seriously doubted that in several other parts of the globe aggression is only deterred by the avowed or understood policy of a powerful opponent, not by the strength of the present possessor. This is the significance of the new Anglo-Japanese agreement, and also of the more venerable Monroe Doctrine of the United States, though that is applicable in another quarter. The parties to either of these policies is interested in the success of the other.

It seems demonstrable, therefore, that as commerce is the engrossing and predominant interest of the world to-day, so, in consequence of its acquired expansion, oversea commerce, over-sea political acquisition, and maritime commercial routes are now the primary objects of external policy among nations. The instrument for the maintenance of policy directed upon these objects is the Navy of the several States; for, whatever influence we attribute to moral ideas, which I have no wish to undervalue, it is certain that, while right rests upon them for its sanction, it depends upon force for adequate assertion against the too numerous, individuals or communities, who either disregard moral sanctions, or reason amiss concerning them.

Further, it is evident that for the moment neither South America nor Africa is an immediate object of far-reaching commercial ambition, to be compassed by political action. Whatever the future may have in store for them, a variety of incidents have relegated them for the time to a position of secondary interest. Attention has centred upon the Pacific generally, and upon the future of China particularly. The present distribution of navies indicates this; for while largely a matter of tradition and routine, nevertheless the assignment of force follows the changes of political circumstances, and undergoes gradual modifications, which reflect the conscious or unconscious sense of the nation that things are different. It is not insignificant that the preponderant French fleet is now in the Mediterranean, whereas it once was in the Atlantic ports; and memories which stretch a generation back can appreciate the fact and the meaning of the diminution of British force on the east and west coasts of America, as also of the increase of Russian battleship force in China seas. Interests have shifted.

Directly connected with these new centres of interest in the Far East, inseparable from them in fact or in policy, are the commercial routes which lead to them. For the commerce and navies of Europe this route is by the Mediterranean and the Suez Canal. This is the line of communication to the objective of interest. The base of all operation, political or military,—so far as the two are separable,—is in the mother countries. These—the base, the objective, and the communications—are the conditions of the problem by which the distribution of naval force is ultimately to be determined. It is to be remarked, however, that while the dominant factor of the three is the line of communication between base and objective, the precise point or section of this upon which control rests, and on which mobile force must be directed, is not necessarily always the same. The distribution of force must have regard to possible changes of dispositions, as the conditions of a war vary.

Every war has two aspects, the defensive and the offensive, to each of which there is a corresponding factor of activity. There is something to gain, the offensive; there is something to lose, the defensive. The ears of men, especially of the uninstructed, are more readily and sympathetically open to the demands of the latter. It appeals to the conservatism which is dominant in the well-to-do, and to the widespread timidity which hesitates to take any risk for the sake of a probable though uncertain gain. The sentiment is entirely respectable in itself, and more than respectable when its power is exercised against breach of the peace for other than the gravest motives—for any mere lucre of gain. But its limitations must be understood. A sound defensive scheme, sustaining the bases of the national force, is the foundation upon which war rests; but who lays a foundation without intending a superstructure? The offensive element in warfare is the superstructure, the end and aim for which the defensive exists, and apart from which it is to all purposes of war worse than useless.

When war has been accepted as necessary, success means nothing short of victory; and victory must be sought by offensive measures, and by them only can be insured. "Being in, bear it, that the opposer may be ware of thee." No mere defensive attitude or action avails to such end. Whatever the particular mode of offensive action adopted, whether it be direct military attack, or the national exhaustion of the opponent by cutting off the sources of national well-being, whatsoever method may be chosen, offence, injury, weakening of the foe, to annihilation if need be, must be the guiding purpose of the belligerent. Success will certainly attend him who drives his adversary into the position of the defensive and keeps him there.

Offence therefore dominates, but it does not exclude. The necessity for defence remains obligatory, though subordinate. The two are complementary. It is only in the reversal of *rôles*, by which priority of importance is assigned to the defensive, that ultimate defeat is involved. Nor is this all. Though opposed in idea and separable in method of action, circumstances not infrequently have permitted the union of the two in a single general plan of campaign, which protects at the same time that it attacks. "Fitz James's blade was sword and shield." Of this the system of blockades by the British Navy during the Napoleonic wars was a marked example. Thrust up against the ports of France, and lining her coasts, they covered—shielded—the operations of their own commerce and cruisers in every sea; while at the same time, crossing swords, as it were, with the fleets within, ever on guard, ready to attack, should the enemy give an opening by quitting the shelter of his ports, they frustrated his efforts at a combination of his squadrons by which alone he could hope to reverse conditions. All this was defensive; but the same operation cut the sinews of the enemy's power by depriving him of sea-borne commerce, and promoted the reduction of his colonies. Both these were measures of offence; and both, it may be added, were directed upon the national communications, the sources of national well-being. The means was one, the effect twofold.

It is evident also that offensive action depends for energy upon the security of the several places whence its resources are drawn. These are appropriately called "bases," for they are the foundations—more exactly, perhaps, the roots—severed from which vigor yields to paralysis. Still more immediately disastrous would be the destruction or capture of the base itself. Therefore, whether it be the home country in general, the centre of the national power, or the narrower localities where are concentrated the materials of warfare in a particular region, the base, by its need of protection, represents distinctively the defensive element in any campaign. It must be secured at all hazards; though, at the same time, be it clearly said, by recourse to means which shall least fetter the movements of the offensive factor—the mobile force, army or navy. On the other hand, the objective represents with at least equal exclusiveness the offensive element; there, put it at the least, preponderance over the enemy, not yet existent, is to be established by force. The mere effort to get from the base to the objective is an offensive movement; but the ground intervening between the two is of more complex character. Here, on the line of communications, offence and defence blend. Here the belligerent whose precautions secure suitable permanent positions, the defensive element, and to them assign proportionate mobile force, the offensive factor, sufficient by superiority to overpower his opponent, maintains, by so far and insomuch, his freedom and power of action at the distant final objective; for he controls for his own use the indispensable artery through which the national life-blood courses to the distant fleet, and by the same act he closes it to his enemy. Thus again offence and defence meet, each contributing its due share of effect, unified in method and result by an accurate choice of the field of exertion, of that section of the line of communications where power needs to be mainly exerted.

In purely land warfare the relative strength of the opponents manifests itself in the length of the line of communications each permits itself; the distance, that is, which it ventures to advance from its base towards the enemy. The necessary aim of both is superiority at the point of contact, to be maintained either by actual preponderance of numbers, or else by a combination of inferior numbers with advantageous position. The original strength of each evidently affects the distance that he can thus advance, for the line of communication behind him must be secured by part of his forces, because upon it he depends for almost daily supplies. The weaker therefore can go least distance, and may even be compelled to remain behind the home frontier,—a bare defensive,—yielding the other the moral and material advantage of the offensive. But commonly, in land war, each adversary has his own line of communication, which is behind him with respect to his opponent; each being in a somewhat literal sense opposite, as well as opposed, to the other, and the common objective, to be held by the one or carried by the other, lying between them. The strategic aim of both is to menace, or even to sever permanently, the other's communications; for if they are immediately threatened he must retreat, and if sundered he must surrender. Either result is better obtained by this means than by the resort to fighting, for it saves bloodshed, and therefore economizes power for the purpose of further progress.

Maritime war has its analogy to these conditions, but it ordinarily reproduces them with a modification peculiar to itself. In it the belligerents are not usually on opposite sides of the common objective—though they may be so—but proceed towards it by lines that in general direction are parallel, or convergent, and may even be identical. England and France lie side by side, and have waged many maritime wars; but while there have been exceptions, as Gibraltar and Minorca, or when the command of the Channel was in dispute, the general rule has been that the scene of operations was far distant from both, and that both have approached it by substantially the same route. When the prospective theatre of war is reached, the fleet there depends partly upon secondary local bases of supplies, but ultimately upon the home country, which has continually to renew the local deposits, sending stores forward from time to time over the same paths that the fleets themselves travelled. The security of those sea-roads is therefore essential and the dependence of the fleets upon them for supplies of every kind—pre-eminently of coal—reproduces the land problem of communications in a specialized form. The two have to contest the one line of communications vital to both. It becomes therefore itself an objective, and all the more important because the security of military communications entails in equal measure that of the nation's commerce. In broad generalization, the maritime line of communications is the ocean itself, an open plain, limited by no necessary highways, such as the land has to redeem from the obstacles which encumber it, and largely devoid of the advantages of position that the conformation of ground may afford in a shore battlefield. In so far control depends upon superior numbers only, and the give and take which history records, where disparity has not been great, has gone far to falsify the frequent assertion that the ocean acknowledges but one mistress; but as the sea-road draws near a coast, the armed vessels that assail or protect are facilitated in their task if the shore affords them harbors of refuge and supply. A ship that has to go but fifty miles to reach her field of operation will do in the course of a year the work of several ships that have to go five hundred. Fortified naval depots at suitable points therefore increase numerical force by multiplying it, quite as the possession of strategic points, or the lay of the ground of a battlefield, supply numerical deficiencies.

Hence appears the singular strategic—and, because strategic, commercial—interest of a narrow or landlocked sea, which is multiplied manifold when it forms an essential link in an important maritime route. Many widely divergent tracks may be traced on the ocean's unwrinkled brow; but specifically the one military line of communications between any two points of its surface is that which is decisively the shortest. The measure of force between opponents in such a case depends therefore not only upon superiority at the objective point, but upon control of that particular line of communications; for so only can superiority be maintained. The belligerent who, for any disadvantage of numbers, or from inferiority of strength as contrasted with the combined numbers and position of his opponent, cannot sustain his dominant hold there is already worsted.

To this consideration is due the supreme importance of the Mediterranean in the present conditions of the communications and policies of the world. From the commercial point of view it is much the shortest, and therefore the principal, sea route between Europe and the Farther East. At the present time very nearly one-third of the home trade, the exports and imports, of Great Britain originates in or passes through the Mediterranean; and the single port of Marseilles handles a similar proportion of all the sea-borne commerce of France. From the military standpoint, the same fact of shortness, combined with the number and rivalry of national tenures established throughout its area, constitutes it the most vital and critical link in an interior line between two regions of the gravest international concern. In one of these, in Europe, are situated the bases, the home dominions, of the European Powers concerned, and in the other the present chief objective of external interest to all nations of to-day—that Farther East and western Pacific upon which so many events have conspired recently to fasten the anxious attention of the world.

The Mediterranean therefore becomes necessarily the centre around which must revolve the strategic distribution of European navies. It does not follow, indeed, that the distribution of peace reproduces the dispositions for war; but it must look to them, and rest upon the comprehension of them. The decisive point of action in case of war must be recognized and preparation made accordingly; not only by the establishment of suitable positions, which is the naval strategy of peace, but by a distinct relation settled between the numbers and distribution of vessels needed in war and those maintained in peace. The Mediterranean will be either the seat of one dominant control, reaching thence in all directions, owning a single mistress, or it will be the scene of continual struggle. Here offence and defence will meet and blend in their general manifestation of mobile force and fortified stations. Elsewhere the one or the other will have its distinct sphere of predominance. The home waters and their approaches will be the scene of national defence in the strictest and most exclusive sense; but it will be defence that exists for the foundation, upon which reposes the struggle for, or the control of, the Mediterranean. The distant East, in whatever spot there hostilities may rage, will represent, will be, the offensive sphere; but the determination of the result, in case of prolongation of war, will depend upon control of the Mediterranean. In the degree to which that is insured defence will find the test of its adequacy, and offence the measure of its efficiency.

In this combination of the offensive and defensive factors the Mediterranean presents an analogy to the military conditions of insular states, such as Great Britain and Japan, in which the problem of national defence becomes closely identified with offensive action. Security, which is simply defence in its completed result, depends for them upon control of the sea, which can be assured only by the offensive action of the national fleet. Its predominance over that of the enemy is sword and shield. It is a singular advantage to have the national policy in the matter of military development and dispositions so far simplified and unified as it is by this consideration. It much more than compensates for the double line of communications open to a continental state, the two strings to its bow, by its double frontiers of sea and land; for with the two frontiers there is double exposure as well as double utility. They require two-fold protective action, dissipating the energies of the nation by dividing them between two distinct objects, to the injury of both.

An insular state, which alone can be purely maritime, therefore contemplates war from a position of antecedent probable superiority from the two-fold concentration of its policy; defence and offence being closely identified, and energy, if exerted judiciously, being fixed upon the increase of naval force to the clear subordination of that more narrowly styled military. The conditions tend to minimize the division of effort between offensive and defensive purpose, and, by greater comparative development of the fleet, to supply a larger margin of disposable numbers in order to constitute a mobile superiority at a particular point of the general field. Such a decisive local superiority at the critical point of action is the chief end of the military art, alike in tactics and strategy. Hence it is clear that an insular state, if attentive to the conditions that should dictate its policy, is inevitably led to possess a superiority in that particular kind of force, the mobility of which enables it most readily to project its power to the more distant quarters of the earth, and also to change its point of application at will with unequalled rapidity.

The general considerations that have been advanced concern all the great European nations, in so far as they look outside their own continent, and to maritime expansion, for the extension of national influence and power; but the effect upon the action of each differs necessarily according to their several conditions. The problem of sea-defence, for instance, relates primarily to the protection of the national commerce everywhere, and specifically as it draws near the home ports; serious attack upon the coast, or upon the ports themselves, being a secondary consideration, because little likely to befall a nation able to extend its power far enough to sea to protect its merchant ships. From this point of view the position of Germany is embarrassed at once by the fact that she has, as regards the world at large, but one coast-line. To and from this all her sea commerce must go; either passing the English Channel, flanked for three hundred miles by France on the one side and England on the other, or else going north about by the Orkneys, a most inconvenient circuit, and obtaining but imperfect shelter from recourse to this deflected route. Holland, in her ancient wars with England, when the two were fairly matched in point of numbers, had dire experience of this false position, though her navy was little inferior in numbers to that of her opponent. This is another exemplification of the truth that distance is a factor equivalent to a certain number of ships. Sea-defence for Germany, in case of war with France or England, means established naval predominance at least in the North Sea; nor can it be considered complete unless extended through the Channel and as far as Great Britain will have to project hers into the Atlantic. This is Germany's initial disadvantage of position, to be overcome only by adequate superiority of numbers; and it receives little compensation from the security of her Baltic trade, and the facility for closing that sea to her enemies. In fact, Great Britain, whose North Sea trade is but one-fourth of her total, lies to Germany as Ireland does to Great Britain, flanking both routes to the Atlantic; but the great development of the British seacoast, its numerous ports and ample internal communications, strengthen that element of sea-defence which consists in abundant access to harbors of refuge.

For the Baltic Powers, which comprise all the maritime States east of Germany, the commercial drawback of the Orkney route is a little less than for Hamburg and Bremen, in that the exit from the Baltic is nearly equidistant from the north and south extremities of England; nevertheless the excess in distance over the Channel route remains very considerable. The initial naval disadvantage is in no wise diminished. For all the communities east of the Straits of Dover it remains true that in war commerce is paralyzed, and all the resultant consequences of impaired national strength entailed, unless decisive control of the North Sea is established. That effected, there is security for commerce by the northern passage; but this alone is mere defence. Offence, exerted anywhere on the globe, requires a surplusage of force, over that required to hold the North Sea, sufficient to extend and maintain itself west of the British Islands. In case of war with either of the Channel Powers, this means, as between the two opponents, that the eastern belligerent has to guard a long line of communications, and maintain distant positions, against an antagonist resting on a central position, with interior lines, able to strike at choice at either wing of the enemy's extended front. The relation which the English Channel, with its branch the Irish Sea, bears to the North Sea and the Atlantic—that of an interior position—is the same which the Mediterranean bears to the Atlantic and the Indian Sea; nor is it merely fanciful to trace in the passage round the north of Scotland an analogy to that by the Cape of Good Hope. It is a reproduction in miniature. The conditions are similar, the scale different. What the one is to a war whose scene is the north of Europe, the other is to operations by European Powers in Eastern Asia.

To protract such a situation is intolerable to the purse and *morale* of the belligerent who has the disadvantage of position. This of course leads us straight back to the fundamental principles of all naval war, namely, that defence is insured only by offence, and that the one decisive objective of the offensive is the enemy's organized force, his battle-fleet. Therefore, in the event of a war between one of the Channel Powers, and one or more of those to the eastward, the control of the North Sea must be at once decided. For the eastern State it is a matter of obvious immediate necessity, of commercial self-preservation. For the western State the offensive motive is equally imperative; but for Great Britain there is defensive need as well. Her Empire imposes such a development of naval force as makes it economically impracticable to maintain an army as large as those of the Continent. Security against invasion depends therefore upon the fleet. Postponing more distant interests, she must here concentrate an indisputable superiority. It is, however, inconceivable that against any one Power Great Britain should not be able here to exert from the first a preponderance which would effectually cover all her remoter possessions. Only an economical decadence, which would of itself destroy her position among nations, could bring her so to forego the initial advantage she has, in the fact that for her offence and defence meet and are fulfilled in one factor, the command of the sea. History has conclusively demonstrated the inability of a state with even a single continental frontier to compete in naval development with one that is insular, although of smaller population and resources. A coalition of Powers may indeed affect the balance. As a rule, however, a single state against a coalition holds the interior position, the concentrated force; and while calculation should rightly take account of possibilities, it should beware of permitting imagination too free sway in presenting its pictures. Were the eastern Powers to combine they might prevent Great Britain's use of the North Sea for the safe passage of her merchant shipping; but even so she would but lose commercially the whole of a trade, the greater part of which disappears by the mere fact of war. Invasion is not possible, unless her fleet can be wholly disabled from appearing in that sea.

From her geographical position, she still holds her gates open to the outer world, which maintains three-fourths of her commerce in peace.

As Great Britain, however, turns her eyes from the North and Baltic Seas, which in respect to her relations to the world at large may justly be called her rear, she finds conditions confronting her similar to those which position entails upon her eastern neighbors. Here, however, a comparison is to be made. The North Sea is small, its coast-line contracted, the entrance to the Baltic a mere strait. Naval preponderance once established, the lines of transit, especially where they draw near the land, are easily watched. Doubtless, access to the British Islands from the Atlantic, if less confined by geographical surroundings, is constricted by the very necessity of approaching at all; but a preponderant fleet maintained by Great Britain to the south-west, in the prolongation of the Channel, will not only secure merchant shipping within its own cruising-ground, but can extend its support by outlying cruisers over a great area in every direction. A fleet thus in local superiority imposes upon cruisers from the nearest possible enemy—France—a long circuit to reach the northern approaches of the islands, where they will arrive more or less depleted of coal, and in danger from ships of their own class resting on the nearer ports of Scotland or Ireland. Superiority in numbers is here again counterbalanced by advantage of position. Vessels of any other country, south or east, are evidently under still greater drawbacks.

As all the Atlantic routes and Mediterranean trade converge upon the Channel, this must be, as it always has been, among the most important stations of the British Navy. In the general scheme its office is essentially defence. It protects the economical processes which sustain national endurance, and thus secures the foundation on which the vigor of war rests. But its scope must be sanely conceived. Imaginative expectation and imaginative alarms must equally be avoided; for both tend to exaggerate the development of defensive dispositions at the expense of offensive power. Entire immunity for commerce must not be anticipated, nor should an occasional severe blow be allowed to force from panic concessions which calm reason rejects. Inconvenience and injury are to be expected, and must be borne in order that the grasp upon the determining points of war may not be relaxed. It will be the natural policy of an enemy to intensify anxiety about the Channel, to retain or divert thither force which were better placed elsewhere. By the size of her navy and by her geographical situation France is the most formidable maritime enemy of Great Britain, and therefore supplies the test to which British dispositions must be brought; but it is probable that in war, as now in peace, France must keep the larger part of her fleet in the Mediterranean. Since the days of Napoleon she has given hostages to fortune in the acquisition of her possessions on the African continent and beyond Suez. Her position in the Mediterranean has become to her not only a matter of national sentiment, which it long has been, but a question of military importance much greater than when Corsica was all she owned there. It is most unlikely that Brest and Cherbourg combined will in our day regain the relative importance of the former alone, a century ago.

In view of this, and barring the case of a coalition, I conceive that the battle-ships of the British Channel Fleet would not need to outnumber those of France in the near waters by more than enough to keep actually at sea a force equal to hers. A surplus for reliefs would constitute a reserve for superiority; that is all. The great preponderance required is in the cruisers, who are covered in their operations by the battle-fleet; the mere presence of the latter with an adequate scouting system secures them from molestation. Two classes of cruisers are needed, with distinct functions; those which protect commerce by the strong hand and constant movement, and those that keep the battle-fleet informed of the enemy's actions. It is clear that the close watching of hostile ports, an operation strictly tactical, has undergone marked changes of conditions since the old days. The ability to go to sea and steer any course under any conditions of wind, and the possibilities of the torpedo-boat, exaggerated though these probably have been in anticipation, are the two most decisive new factors. To them are to be added the range of coast guns, which keeps scouts at a much greater distance than formerly, and the impossibility now of detecting intentions which once might be inferred from the conditions of masts and sails.

On the other hand the sphere of effectiveness has been immensely increased for the scout by the power to move at will, and latterly by the wireless telegraphy. With high speed and large numbers, it should be possible to sweep the surroundings of any port so thoroughly as to make the chance of undetected escape very small, while the transmission of the essential facts—the enemy's force and the direction taken—is even more certain than detection. A lookout ship to-day will not see an enemy going off south with a fresh fair breeze, which is for herself a head wind to reach her own fleet a hundred miles to the northward. She may not need even to steam to the main body; but, telephoning the news, she will seek to keep the enemy in sight, gathering round her for the same work all of her own class within reach of her electric voice. True, an enemy may double on his track, or otherwise ultimately elude; but the test so imposed on military sagacity and inference is no greater than it formerly was. The data are different; the problem of the same class. Where can he go fruitfully? A raid? Well, a raid, above all a maritime raid, is only a raid; a black eye, if you will, but not a bullet in the heart, nor yet a broken leg. To join another fleet? That is sound, and demands action; but the British battle-fleet having immediate notice, and a fair probability of more information, should not be long behind. There is at all events no perplexity exceeding that with which men of former times dealt successfully. In the same way, and by the same methods, it should be possible to cover an extensive circumference to seaward so effectively that a merchant vessel reaching any point thereof would be substantially secure up to the home port.

The battle-fleet would be the tactical centre upon which both systems of scouts would rest. To close-watch a port to-day requires vessels swifter than the battle-ships within, and stronger in the aggregate than their cruiser force. The former then cannot overtake to capture, nor outrun to elude; and the latter, which may overtake, cannot drive off their post, nor successfully fight, because inferior in strength. Add to the qualities thus defined sufficient numbers to watch by night the arc of a circle of five miles radius, of which the port is the centre, and you have dispositions extremely effective against an enemy's getting away unperceived. The vessels nearest in are individually so small that the loss of one by torpedo is militarily immaterial; moreover, the chances will by no means all be with the torpedo-boat. The battle-fleet, a hundred or two miles distant it may be, and in a different position every night, is as safe from torpedo attack as ingenuity can place it. Between it and the inside scouts are the armored cruisers, faster than the hostile battle-fleet, stronger than the hostile cruisers. These are tactical dispositions fit for to-day; and in essence they reproduce those of St. Vincent before Brest, and his placing of Nelson at Cadiz with an inshore squadron, a century ago. "A squadron of frigates and cutters plying day and night in the opening of the Goulet; five ships-of-the-line anchored about ten miles outside; and outside of them again three of-the-line under sail." The main body, the battle-fleet of that time, was from twenty-five to forty miles distant,—the equivalent in time of not less than a hundred miles to-day.

Keeping in consideration these same waters, the office and function of the Channel Fleet may be better realized by regarding the battleships as the centre, from which depart the dispositions for watching, not only the enemy's port, but also the huge area to seaward which it is desired to patrol efficiently for the security of the national commerce. Take a radius of two hundred miles; to it corresponds a semicircle of six hundred, all within Marconi range of the centre. The battle-fleet never separates. On the far circumference move the lighter and swifter cruisers; those least able to resist, if surprised by an enemy, but also the best able to escape, and the loss of one of which is inconsiderable, as of the inner cruisers off the port. Between them and the fleet are the heavier cruisers, somewhat dispersed, in very open order, but in mutual touch, with a squadron organization and a plan of concentration, if by mischance an enemy's division come upon one of them unawares. Let us suppose, under such a danger, they are one hundred miles from the central body. It moves out at twelve miles an hour, they in at fifteen. Within four hours the force is united, save the light cruisers. These, as in all ages, must in large measure look out for themselves, and can do so very well.

Granting, as required by the hypothesis, equality in battleships and a large preponderance in cruisers,—not an unreasonable demand upon an insular state,—it seems to me that for an essentially defensive function there is here a fairly reliable, systematized, working disposition. It provides a semi-circumference of six hundred miles, upon reaching any point of which a merchant ship is secure for the rest of her homeward journey. While maintained, the national frontier is by so much advanced, and the area of greatest exposure for the merchant fleet equally reduced. Outside this, cruising as formerly practised can extend very far a protection, which, if less in degree, is still considerable. For this purpose, in my own judgment, and I think by the verdict of history, dissociated single ships are less efficient than cruiser-squadrons, such as were illustrated by the deeds of Jean Bart and Pellew. One such, a half-dozen strong, west of Finisterre, and another west of Scotland, each under a competent chief authorized to move at discretion over a fairly wide area, beyond the bailiwick of the commander-in-chief, would keep enemies at a respectful distance from much more ground than he actually occupies; for it is to be remembered that the opponent's imagination of danger is as fruitful as one's own.

In conception, this scheme is purely defensive. Incidentally, if opportunity offer to injure the enemy it will of course be embraced, but the controlling object is to remove the danger to home commerce by neutralizing the enemy's fleet. To this end numbers and force are calculated. This done, the next step is to consider the Mediterranean from the obvious and inevitable military point of view that it is the one and only central position, the assured control of which gives an interior line of operations from the western coast of Europe to the eastern waters of Asia. To have assured safety to the home seas and seaboard is little, except as a means to further action; for, if to build without a foundation is disastrous, to lay foundations and not to be able to build is impotent, and that is the case where disproportioned care is given to mere defensive arrangements. The power secured and stored at home must be continually transmitted to the distant scene of operations, here assumed, on account of the known conditions of world politics, to be the western Pacific, which, under varying local designations, washes the shores of the Farther East.

It has been said that in the Mediterranean, as the principal link in the long chain of communications, defence and offence blend. Moreover, since control here means assured quickest transmission of reinforcements and supplies in either direction, it follows that, while preponderance in battle-ship force is essential in the Far East, where if war occurs the operations will be offensive, such predominance in the Mediterranean, equally essential in kind, must be much greater in degree. In fact, the offensive fleet in the Eastern Seas and the defensive fleet in the Channel are the two wings, or flanks, of a long front of operations, the due security of both of which depends upon the assured tenure of the central position. Naturally, therefore, the Mediterranean fleet, having to support both, possibly even to detach hurriedly to one or the other, has in itself that combination of defensive and offensive character which ordinarily inheres in sea communications as such.

If this assertion be accepted in general statement, it will be fortified by a brief consideration of permanent conditions; with which it is further essential to associate as present temporary factors the existing alliances between France and Russia, Great Britain and Japan. The Triple Alliance, of the renewal of which we are assured, does not contemplate among its objects any one that is directly affected by the control of the Mediterranean. Should an individual member engage in war having its scene there, it would be as a power untrammelled by this previous engagement.

History and physical conformation have constituted unique strategic conditions in the Mediterranean. To history is due the existing tenure of positions, the bases, of varying intrinsic value, and held with varying degrees of power and firmness by several nations in several quarters. To examine these minutely and weigh their respective values as an element of strategic effect would be indeed essential to the particular planning of a naval campaign, or to the proper determination of the distribution of naval force, with a view to the combinations open to one's self or the enemy; but a paper dealing with general conditions may leave such detailed considerations to those immediately concerned. It must be sufficient to note the eminently central position of Malta, the unique position of Gibraltar, and the excentric situation of Toulon relatively to the great trade route. By conformation the Mediterranean has, besides the artificial canal,—the frailest and most doubtful part of the chain,—at least three straits of the utmost decisive importance, because there is to them no alternative passage by which vessels can leave the sea, or move from one part of it to another. In the Caribbean Sea, which is a kind of Mediterranean, the multiplicity of islands and passages reduces many of them to inconsequence, and qualifies markedly the effect of even the most important; but, in the Mediterranean, the Dardanelles, Gibraltar, and the belt of water separating the toe of Italy from Cape Bon in Africa, constitute three points of transit which cannot be evaded. It is true that in the last the situation of the island of Sicily allows vessels to go on its either side; but the surrounding conditions are such that it is scarcely possible for a fleet to pass undetected by an adversary making due use of his scouts. These physical peculiarities, conjointly with the positions specified, are the permanent features, which must underlie and control all strategic plans of Mediterranean Powers, among whom Russia must be inferentially included.

Geographically, Great Britain is an intruder in the Mediterranean. Her presence there at all, in territorial tenure, is distinctively military. This is witnessed also by the character of her particular possessions. Nowhere does the vital energy of sea power appear more conspicuously, as self-expansive and self-dependent. To its historical manifestation is due the acquisitions which make the strength of her present position; but, as in history, so now, sea power itself must continue to sustain that which it begat. The habitual distribution of the warships of the United Kingdom must provide for a decisive predominance here, upon occasion arising, over any probable combination of enemies. Such provision has to take account not only of the total force of hostile divisions within and without the Mediterranean, but of movements intended to transfer one or more from or to that sea from other scenes of operations. Prevention of these attempts is a question, not of numbers chiefly, but of position, of stations assigned, of distribution. Predominance, to be militarily effectual, means not only an aggregate superiority to the enemy united, but ability to frustrate, before accomplishment, concentrations which might give him a local superiority anywhere. This is a question of positions more even than of numbers. In the Mediterranean, as the great centre, these two factors must receive such mutual adjustment as shall outweigh the combination of them on the part of the adversary. Where one is defective the other must be increased. The need is the more emphatic when the nation itself is external and distant from the sea, while possible antagonists, as Russia and France, are territorially contiguous; for it can scarcely be expected that the Russian Black Sea fleet would not force its way through the Dardanelles upon urgent occasion.

Evidently, too, Japan cannot in the near future contribute directly to maintain Great Britain in the Mediterranean. On the contrary, the declarations of Russia and France make plain that, if war arise, Japan must be supported in the Far East by her ally against a coalition, the uncertain element of which is the force that France will feel able to spare from her scattered, exposed interests. Russia labors under no such distraction; her singleness of eye is shown by the fact that the more efficient, and by far the larger part, of her so-called Baltic fleet is now in the waters of China. In numbers and force she has there a substantial naval equality with Japan, but under a disadvantage of position like that of Great Britain in the Mediterranean, in being remote from the centre of her power, imperfectly based, as yet, upon local resources, and with home communications by the shortest route gravely uncertain. Under these circumstances the decided step she has taken in the reinforcement of her Eastern Navy, carries the political inference that she for the present means to seek her desired access to unfrozen waters in Eastern Asia, preferably to the Mediterranean or the Persian Gulf. Having in view local difficulties and antagonistic interests elsewhere, this conclusion was probably inevitable; but its evident acceptance is notable.

For Great Britain it is also most opportune; and this raises a further question, attractive to speculative minds, viz.:

whether the Anglo-Japanese agreement has had upon Russia a stimulating or a deterrent effect? If it has increased her determination to utilize her present advantages, as represented in Port Arthur and its railroad, it would be in the direct line of a sound British policy; for it fixes the reasonable satisfaction of Russia's indisputable needs in a region remote from the greater interests of Great Britain, yet where attempts at undue predominance will elicit the active resistance of many competitors, intent upon their own equally indisputable rights. The gathering of the eagles on the coasts of China is manifest to the dullest eye. But should the alliance have the contrary effect of checking Russian development in that direction, her irrepressible tendency to the sea is necessarily thrown upon a quarter—the Levant or Persia—more distinctly ominous, and where, in the last named at least, Great Britain would find no natural supporter, enlisted by similarity of interest. The concentration of Russian ships in the East, taken in connection with the general trend of events there, is, however, as clear an indication of policy as can well be given.

In connection with the substantial numerical equality of Japan and Russia is to be taken, as one of the ascertained existing conditions, instituted so recently as to have a possible political significance, the reorganization of the French divisions beyond Suez into a single command, and the numbers thereto assigned. It is not to be supposed that this new disposition has been adopted without consideration of the new combinations indicated by the Anglo-Japanese treaty. It may even be in direct consequence. The relative strengths of this extensive eastern command and of the French Mediterranean fleet should in close measure reflect the official consciousness of the general naval situation, and of the power of France to give support to her recognized ally; directly in the East, and indirectly by military influence exerted upon the Mediterranean. Supposing Great Britain, on the other hand, to have made provision for the defensive control of the approaches to her home ports, how will she, and how can she, assure the joint ascendancy of herself and her ally in the Farther East, the scene of the offensive, and her own single preponderance in the Mediterranean, the main link in the communications? These are the two intricate factors for consideration, calling for plans and movements not primarily defensive but offensive in scope. For France and for Great Britain, as a party to an alliance, the question is urgent, "How far can I go, how much spare from the Mediterranean to the East? In assisting my ally there, unless I bring him predominance, or at least nearly an equality, I waste my substance, little helping him. If paralyzed in the Mediterranean, thrown on a mere defensive, my force in the East is practically cut off. Like a besieged garrison, it may endure till relieved; but the situation is critical while it lasts, and carries imminent possibilities of disaster."

In approaching a military subject of this character it is necessary first and for all to disabuse the mind of the idea that a scheme can be devised, a disposition imagined, by which all risk is eliminated. Such an attractive condition of absolute security, if realized, would eliminate all war along with its risks. A British distribution, most proper for the Mediterranean alone, may entail the danger that a hostile body may escape into the Atlantic, may unite with the Brest and Cherbourg divisions against the Channel Fleet, and overwhelm the latter. True; but imagination must work both ways. It may also be that the escape cannot but be known at Gibraltar, telegraphed to England, and the fleet warned betimes so that the reserve ships, which give it a superiority to either detachment of the enemy, might join, and that its scouts, stationed as previously suggested, would gain for it the two hours of time needed to deal decisively with one division before the other turns up. These probabilities, known to the enemy, affect his actions just as one's own risks move one's self. Listen to Nelson contemplating just this contingency. "If the Ferrol squadron joins the Toulon, they will much outnumber us, but in that case I shall never lose sight of them, and Pellew" (from before Ferrol) "will soon be after them." But he adds, confirmatory of the need of numerous scouts, then as now, "I at this moment want ten frigates or sloops, when I believe neither the Ferrol or Toulon squadron could escape me." By this, I understand, is clearly intimated that he could look out both ways, intercept the first comer, frustrate the junction, and beat them in detail. If not before the action, Pellew would arrive in time to repair Nelson's losses and restore equality. The change in modern conditions would favor the modern Pellew more than the adversary.

So again disturbing political possibilities must be reasonably viewed. It may be that the whole Continent not only dislikes Great Britain, but would willingly combine for her military destruction; and that, if war begin, such a combination may come to pass. It may be; but this at least is certain, that interest, not liking, will decide so grave a matter. In the calculation of final issues, of national expenditure, of profit and loss, of relative national predominance resulting from a supposed success, I incline to think that Imperial Federation will be a far less difficult achievement than framing such a coalition. If the two dual alliances, the mutual opposition of which is apparent, come to blows, Germany may see it to her interest to strike hands with Russia and France; but it seems to me it would be so much more her interest to let them exhaust themselves, to the relief of her two flanks, that I find it difficult to believe she would not herself so view the question. There is one qualifying consideration. Germany cannot but wish a modification in the effect exerted upon her maritime routes by the position of Great Britain, already noted. As geographical situation cannot be changed, the only modification possible is the decrease of Great Britain's power by the lessening of her fleet. But, grant that object gained by such coalition, what remains? A Channel dominated by the French Navy no longer checked by the British; whereas with the latter as an ally the Channel would be almost as safe as the Kiel canal. If this remark is sound, it is but an illustration of the choice of difficulties presented by attempts to change permanent conditions by artificial combinations. As a matter of fact, no single power in Europe, save possibly Russia, is individually so weighty as to see without apprehension the effective elimination of any one factor in the present balance of power. The combined position and numbers of Russia do give her a great defensive security in her present tenures.

Admitting the Mediterranean to be a distinctively and preeminently the crucial feature in any strategic scheme that contemplates Europe and the Farther East as the chief factors of interest, the positions before enumerated, in conjunction with the relative forces of the fleets, constitute the initial strategic situation. Assuming, as is very possible, that the decisive predominance, local or general, desired by either party, does not yet exist, the attempt of each must be to reach some preponderance by playing the game of war; by such applied pressure or strategic movements as shall procure a decisive momentary preponderance in some quarter, the due use of which, by the injury done the enemy, shall establish a permanent and decisive superiority. This is the one object of war scientifically—or better, artistically—considered. The nation that begins with the stronger fleet should initiate some offensive action, with the object of compelling the enemy to fight. This the latter cannot do, unless already in adequate strength at some one point, except by undertaking to combine his divided forces so as to effect a concentration in some quarter. The movements necessary to accomplish this are the opportunity of the offensive, to strike the converging divisions before their junction gives the desired local superiority. Herein is the skill; herein also the chance, the unexpected, the risk, which the best authorities tell us are inseparable from war, and constitute much of its opportunity as of its danger.

How shall the superior fleet exercise the needed compulsion? Ships cannot invade territory, unless there be unprotected navigable rivers. The stronger navy therefore cannot carry war beyond the sea-coast, home to the heart of the enemy, unless indeed its nation in addition to controlling the sea, can transport an overpowering force of troops. Of this the Transvaal war offers an illustration. Possibly, a disabling blow to the British fleet by the navy of one of the great continental armies might present a somewhat similar instance; but when the British fleet is thus enfeebled, Great Britain will be exposed to the conditions which it must be her own first effort, with her supreme navy, to impose on an opponent. Under such circumstances, there will be no need for an enemy to land an invading host on British soil. The interception of commerce at a half-dozen of the principal ports will do the work as surely, if less directly. Similarly, while the British Navy is what it is, the destruction of an enemy's commerce, not only by scattered cruisers at sea, but by a systematized, coherent effort directed against his ports and coasts, both home and colonial, must be the means of inflicting such distress and loss as shall compel his fleet to fight; or, if it still refuse, shall sap endurance by suffering and extenuation.

To effect this requires a battle-fleet superior in the aggregate to the one immediately opposed to it by at least so many ships as shall suffice to allow a constant system of reliefs. The battle-fleet is the solid nucleus of power. From it radiates the system of cruisers by which the trade blockade is maintained in technical, and as far as may be, in actual, efficiency. In case of hostilities with France, for example, the blockade of a principal commercial port, like Havre or Marseille, may be sustained in local efficiency by cruisers; but the security of these, and consequently the maintenance of the blockade, will depend upon such proximity of the battle-fleet as will prevent the French divisions at Cherbourg, Brest, or Toulon, from attacking them, except at great risk of being compelled to an engagement which it is presumably the specific aim of the

British fleet to force. "Not blockade but battle is my aim," said Nelson: "on the sea alone we hope to realize the hopes and expectations of our country." A successful battle in any one quarter clears up the whole situation; that is, in proportion to the results obtained. This qualification is always to be borne in mind by a victorious admiral; for the general relief to his nation will correspond to the use made by him of the particular advantage gained. More or fewer of his ships will be liberated from their previous tasks, and can reinforce the station where the most assured predominance is desired. This by our analysis is the Mediterranean.

History has more than once shown how severe a compulsion may be exerted over an extensive coast by proper dispositions. Where a formidable, though inferior, navy lies in the ports of the blockaded state, the position and management of the battle-fleet, on either side, is the critical military problem. The task of the cruisers is simple, if arduous; to keep near the port assigned them, to hold their ground against equals, to escape capture by superior force. The battle-fleet must be so placed as effectually to cover the cruisers from the enemy's fleet, without unduly exposing itself; above all to torpedo attack. It must be on hand, not only to fight, but to chase to advantage, to make strategic movements, perhaps extensive in range, at short notice. War is a business of positions. Its position, suitably chosen, by supporting the cruiser force, covers the approaches of the national commerce, and also maintains both the commercial blockade and the close watch of the military ports. It may be noted that the commercial blockade is offensive in design, to injure the enemy and compel him to fight, while the other specified functions of the vessels are defensive. We therefore have here again a combination of the two purposes in a single disposition.

For some time to come nations distinctively European must depend upon the Mediterranean as their principal military route to the Far East. In the present condition of the Siberian railroad, Russia shares this common lot. While the other States have no land route whatever, hers is still so imperfect as not to constitute a valid substitute. Moreover, whatever resources of moderate bulk may be locally accumulated,—coal, provisions, ammunition, and stores of various kinds,—reinforcements of vessels, or reliefs to ships disabled by service or in battle can go only by sea. Guns beyond a certain calibre are in like case. Every consideration emphasizes the importance of the Mediterranean. To it the Red Sea is simply an annex, the military status of which will be determined by that of its greater neighbor, qualified in some measure by the tenure of Egypt and Aden.

On the farther side of the isthmus the naval operations throughout Eastern seas will depend for sustained vigor upon contact militarily maintained with the Mediterranean, and through that with home. In these days of cables, the decisive importance of Malta to India, recognized by Nelson and his contemporaries, is affirmed with quadruple force of the sea in which Malta is perhaps the most conspicuously important naval position. Reinforcements sent by the Cape, whether west or east, can always be anticipated at either end of the road by the Power which holds the interior line.

As regards special dispositions for the Eastern seas, embracing under that name all from Suez to Japan, the same factors—numbers and position—dictate distribution. To a central position, if such there be, must be assigned numbers adequate to immediate superiority, in order to control commercial routes, and to operate against the enemy whose approximate force and position are known. Such assignment keeps in view, necessarily, the possibilities of receiving reinforcements from the Mediterranean, or having to send them to China. Ceylon, for example, if otherwise suitable, is nearly midway between Suez and Hong-Kong; in round numbers, 3000 miles from each. Such a position favors a force of battleships as an advanced squadron from the Mediterranean, and would be a provision against a mishap at the canal interrupting reinforcements eastward. Position, with its two functions of distance and resources; there is nothing more prominent than these in Napoleon's analysis of a military situation. Numbers go, as it were, without saying. Where the power was his he multiplied them; but he always remembered that position multiplies spontaneously. He who has but half-way to go does double work. This is the privilege of central position.

The question of the Eastern seas introduces naturally the consideration of what the great self-governing colonies can do, not only for their own immediate security, and that of their trade, but for the general fabric of Imperial naval action, in the coherence of which they will find far greater assurance than in merely local effort. The prime naval considerations for them are that the British Channel Fleet should adequately protect the commerce and shores of the British Islands, and that the Mediterranean Fleet should insure uninterrupted transit for trade and for reinforcements. These effected and maintained, there will be no danger to their territory; and little to their trade except from single cruisers, which will have a precarious subsistence as compared with their own, based upon large self-supporting political communities. Australasia, however, can undoubtedly supply a very important factor, that will go far to fortify the whole British position in the Far East. A continent in itself, with a thriving population, and willing, apparently, to contribute to the general naval welfare, let it frame its schemes and base its estimates on sound lines, both naval and imperial; naval, by allowing due weight to battle force; imperial, by contemplating the whole, and recognizing that local safety is not always best found in local precaution. There is a military sense, in which it is true that he who loses his life shall save it.

In the Eastern seas, Australia and China mark the extremities of two long lines, the junction of which is near India; let us say, for sake of specificness, Ceylon. They are offshoots, each, of one branch, the root of which under present conditions, is the English Channel, and the trunk the Mediterranean. Now it is the nature of extremities to be exposed. To this our feet, hands, and ears bear witness, as does the military aphorism about salients; but while local protection has its value in these several cases, the general vigor and sustenance of the organism as a whole is the truer dependence. To apply this simile: it appears to me that the waters from Suez eastward should be regarded as a military whole, vitally connected with the system to the westward, but liable to temporary interruption at the Canal, against which precaution must be had. This recognizes at once the usual dependence upon the Channel and the Mediterranean, and the coincident necessity of providing for independent existence on emergency. In the nature of things there must be a big detachment east of Suez; the chance of its being momentarily cut off there is not so bad as its being stalled on the other side, dependent on the Cape route to reach the scene. But for the same reason that the Mediterranean and Malta are strategically eminent, because central, (as is likewise the Channel with reference to the North Sea and Atlantic), the permanent strategic centre of the Eastern seas is not by position in China, nor yet in Australia. It is to be found rather at a point which, approximately equidistant from both, is also equidistant from the Mediterranean and the East. Permanent, I say; not as ignoring that the force which there finds its centre may have to remove, and long to remain, at one extremity or another of the many radii thence issuing, but because there it is best placed to move in the shortest time in any one of the several directions. That from the same centre it best protects the general commercial interests is evident from an examination of the maps and of commercial returns.

Whether the essential unity of scope in naval action east of Suez should receive recognition by embracing Australia,

China, and India, under one general command, with local subordinates, is a question administrative as well as strategic. As military policy it has a good side; for commanders previously independent do not always accept ungrudgingly the intrusion of a superior because of emergency of war. Military sensitiveness cannot prudently be left out of calculations. There would be benefit also in emphasizing in public consciousness the essential unity of military considerations, which should dominate the dispositions of the fleet. Non-professional—and even military—minds need the habit of regarding local and general interests in their true relations and proportions. Unless such correct appreciation exist, it is hard to silence the clamor for a simple local security, which is apparent but not real, because founded on a subdivision and dissemination of force essentially contrary to sound military principle. What Australasia needs is not her petty fraction of the Imperial navy, a squadron assigned to her in perpetual presence, but an organization of naval force which constitutes a firm grasp of the universal naval situation. Thus danger is kept remote; but, if it should approach, there is insured within reaching distance an adequate force to repel it betimes. There may, however, be fairly demanded the guarantee for the fleet's action, in a development of local dock-yard facilities and other resources which shall insure its maintenance in full efficiency if it have to come.

In this essential principle other colonies should acquiesce. The essence of the matter is that local security does not necessarily, nor usually, depend upon the constant local presence of a protector, ship or squadron, but upon general dispositions. As was said to and of Rodney, "Unless men take the great line, as you do, and consider the King's whole dominions as under their care, the enemy must find us unprepared somewhere. It is impossible to have a superior fleet in every part."

It is impossible; and it is unnecessary, granting the aggregate superiority at which Great Britain now aims. In the question of the disposition of force three principal elements are distinguishable in the permanent factors which we classify under the general head of "position." These are, the recognition of central positions, of interior lines—which means, briefly, shorter lines—and provision of abundant local dockyard equipment in its widest sense. These furnish the broad outline, the skeleton of the arrangement. They constitute, so to say, the qualitative result of the analysis which underlies the whole calculation. Add to it the quantitative estimate of the interests at stake, the dangers at hand, the advantages of position, in the several quarters, and you reach the assignment of numbers, which shall make the dry bones live with all the energy of flesh and blood in a healthy body; where each member is supported, not by a local congestion of vitality, but by the vigor of the central organs which circulate nourishment to each in proportion to its needs.

CHAPTER IX

THE PERSIAN GULF AND INTERNATIONAL RELATIONS

THE American whom above all others his countrymen delight to honor, more even to-day than a century ago, as his sober wisdom and unselfish patriotism stand in stronger relief on the clear horizon of the past, when he took leave of public life, cautioned his fellow-citizens of that day against "permanent inveterate antipathies against particular nations." In uttering this warning, to which he added certain obvious corollaries as to the effect of prejudice, sympathetic as well as antipathetic, upon action, Washington had vividly in mind American conditions, both present and past, of which he had had bitter official experience. His own people had then divided, and was still farther dividing, in sentiment and utterance, upon lines of sympathy for and against Great Britain and France. Impassioned feeling and fervent speech were doing the deadly work he deplored, in setting man against man, and to some extent section against section, upon issues which were at least not purely of American interest. Harmful at any time, such an opposition of misplaced emotions was peculiarly dangerous then, when the still recent union under the Constitution of 1789 had not yet had time to obliterate the colonial habits of thought, to which the common term "American" loomed far less large, and was far less dear, than

National Review, vol. 40 (September 1902), 27–45. Reprinted in *Retrospect and Prospect* (Boston, 1902), 209–51.

the local appellations of the several States. This inspired Washington's further very serious and, to use his own word, "affectionate" counsels against the spirit of faction and disunion, which, though not confined to our political community, presented special perils to one but lately organized.

Nor was it only against immediate instances of inveterate national antipathies that Washington uttered his warning. These served him merely as pointed illustrations. He based his counsels, as advice to be sound must ever be based, upon permanent general principles. International relations, he said, were not determined, and should not be determined, by sympathy, but by justice and by interest. Justice of course first. However onerous and unsatisfactory, "let existing engagements be observed in their genuine sense." Beyond this, "keep constantly in view that 'tis folly in one nation to look for disinterested favors from another; that it must pay with a portion of its independence for whatever it may accept under that character; that by acceptance it may place itself in the condition of having given equivalent for nominal favors, and yet of being reproached with ingratitude for not giving more."

Here again, in this slightly veiled allusion to the French alliance, was indicated the intrusion of bias into international relations. The help extended by France to the American struggle for independence was indeed real; but as a favor, though given that coloring, it was purely nominal. Yet upon it, so regarded, were based extravagant claims, not only for American sympathy, but for American active support in the early days of the French Revolution. Sight was lost of the notorious fact, that, however disinterested the action of individual Frenchmen, the French government, with proper regard to the interests of its own nation, had simply utilized the revolt of the colonies to renew its old struggle with Great Britain under favorable conditions. A large number of Americans, treasuring the then recent occasions of bitter hostility to Great Britain, responded vehemently; another numerous party, alienated by republican excesses in France, and seeing a truer ideal of liberty in British institutions, recoiled with equal vigor. At a moment when every consideration of expediency dictated political detachment, to the intensification of national life, by pruning superfluous activities and concentrating vital force upon internal consolidation and development, a vast motive power of passion and prejudice was aroused, misdirecting national energy into channels where it not merely ran to waste but corroded the foundations of the Union. On one side and the other, the ideals of national duty and policy became confused with the names of foreign peoples, leading to a bitterness of antagonism that prolonged through a generation the immaturity of the affection uniting the States; maintaining an internal weakness which manifested itself recurrently with each fresh cause of variance, and entailed continued feebleness of external influence until it disappeared forever in the agonies of civil war.

It will doubtless be argued that there is now general recognition that reasoned interest, controlled by justice, is the true regulator of state policy. Possibly; but does practice coincide? Is national calmness or harmony undisturbed, national force unweakened, by sympathies and antipathies which, however otherwise justified, have no proper place in perturbing international conduct? The fostering of an internal spirit of faction is not the only evil effect on national judgment that may arise from extra-national repulsions or attractions. The immediate evil of disruption, which then threatened the United States, is indeed not imminent for political communities of longstanding consolidation; but even into them prepossession indulged for or against other peoples, as such, introduces a motive which is to national efficiency what a morbid growth is to the health of the body. The functions are vitiated, vision impaired, and movement undecided or misdirected; perhaps both. A tendency arises to seek the solution of difficulties in artificial and sometimes complicated international arrangements, contemplating an indefinite future, instead of in simple national procedure meeting each new situation as it develops, governed by a settled general national policy. The latter course may at times incur the reproach of inconsistency through the inevitable necessity of conforming particular measures to unforeseen emergencies; but it may none the less remain most truly consistent in its fixed regard to a few evident leading conditions for which permanency may be predicated. Washington, a man wise with the wisdom that comes of observation in practical life, phrased this for his countrymen, in the connection already quoted, in the words, "Consulting the natural course of things, forcing nothing;" or, as an American experienced in political campaigning once said to me, "Never contrive an opportunity."

Nothing is more fruitful of that frequent charge of bad faith among nations than the attempt to substitute the artificial for the natural. When subsequent experience shows that interest has been elaborately sacrificed because imperfectly comprehended or wholly misunderstood, popular revulsion ultimately exerts over rulers an influence that is compulsive in proportion to the urgency of the situation. It does not follow from this that a nation, as such, has premeditated bad faith, or wilfully accepts it. Nations are not cynical, though individual statesmen have been. There need be no attempt to justify breach of engagement; but it is a very partial view of facts not to recognize that the greater fault lies with those who made a situation which could not be perpetuated, because contrary to the nature of things. Such action should be accepted as a warning that international arrangements can be regarded as sound only when they conform to substantial conditions, relatively at least permanent. If this caution be observed, national policy may through long periods be as enduring as national characteristics admittedly are. National character abides, though nations under impulse are often inconstant. So may national policy, though on occasion fluctuating, or even vacillating, be really constant; but to be so it must conform to the nature of things, consulting—not resisting—their course.

If this be so as regards general policy, it follows that successive questions, as they arise, should be viewed in their relation to that general policy, which it must be assumed is consciously realized in its broad outlines by the governments of the day. Of such questions the prospective status of Persia and the Persian Gulf now forms one, in the consideration of two or three of the great world powers. In their regard to it, and to the various interests or enterprises centring around it, how far are they guided by the natural tendency of things? How far are they seeking to interject artificial arrangements, forced ambitions? What is to be said, from this point of view, of the proposed activities, the various theories of action, suggested political compromises, that here find their origin? As the phrase "world politics" more and more expresses a reality of these latter days, the more necessary does it become to consider each of the several centres of interest as not separate, but having relations to the whole; as contributory to a general balance of constitution, to the health of which it is essential to work according to nature, not contrary to it.

In the general economy of the world, irrespective of political tenures, present or possible, the Persian Gulf is one terminus of a prospective interoceanic railroad. The track of this, as determined by topographical considerations, will take in great part a course over which, at one period and another of history, commerce between the East and West has travelled. Though itself artificial, it will follow a road so far conforming to the nature of things that it has earned in the past the name of the Highway of Nations. The railroad will be one link, as the Persian Gulf is another, in a chain of communication between East and West, alternative to the all-water route by the Suez Canal and the Red Sea. This new line will have over the one now existing the advantage, which rail travel always has over that by water, of greater specific rapidity. It will therefore serve particularly for the transport of passengers, mails, and lighter freights. On the other hand, for bulk of transport, meaning thereby not merely articles singly of great weight or size, but the aggregate amounts of freight that can be carried in a given time, water will always possess an immense and irreversible advantage over land transport for equal distances. This follows directly from the fact that a railroad is essentially narrow. Even with four tracks, it admits of but two trains proceeding abreast in the same direction; whereas natural water ways as a rule permit ships, individually of greater capacity than any single train, to go forward in numbers practically unlimited. A water route is, as it were, a road with numberless tracks. For these reasons, and on account of the first cost of construction, water transport has a lasting comparative cheapness, which so far as can be foreseen will secure to it forever a commercial superiority over that by land. It is also, for large quantities, much more rapid; for, though a train can carry its proper load faster than a vessel can, the closely restricted number of trains that can proceed at once, as compared to the numerous vessels, enables the latter in a given time, practically simultaneously, to deliver a bulk of material utterly beyond the power of the road.

Commercially, therefore, the railroad system, or systems, and their branches, which shall find their terminus at the Persian Gulf, begin at a great disadvantage towards the Suez route, considered as a line of commercial communication between two seas, or between the two continents, Asia and Europe. This, the broad general result, is, however, only one aspect of the relations to world politics. A railroad, as all know, develops the country through which it passes. This means that it there increases existing interests, and creates new ones. Of these it, and through it its owners, become the fostering and controlling centre. Because of this effect, railroads possess a marked local commercial influence; and commercial influence, especially in these days, and in regions where government is weak or remiss, readily becomes political. It is in measure compelled to political action, to protect its varied interests. Furthermore, railroads serve to expedite not only the movement of commerce but the movement of troops. They have therefore military significance, as well as commercial and political. This is a commonplace, upon which it is needless to insist beyond recalling that it inheres in all railroads as such, and therefore in the one under consideration. Finally, while all parts of a commercial route, by land or by sea, have a certain value, supreme importance is accumulated at the termini, the points of arrival or of departure. The operations of commerce,—receipt, distribution, or transshipment,—are there multiplied many fold. This concentration makes them singularly the objects of forcible interference, and consequently attributes to them an importance which is military or naval, according to the locality. This at present is the particular bearing of the Persian Gulf upon world politics. It is closely analogous to that of Port Arthur, which has preceded it so shortly as not yet to be fairly out of sight, as a matter of international heartburnings. Upon the control of it will rest the functioning of the prospective railroad itself, regarded either as a through line of communication, or as a maintainer of local industries by the access it affords them to wider markets. Not only the prosperity of the railroad itself is at stake. The commercial interests that depend upon it, those of the country through which it runs and to which it immediately ministers, and those of many other regions, as producers or consumers, are involved in the political and military status of the Persian Gulf.

Whose affair then is this, intrinsically so important? Not that of all the world, for though all the world may be interested, more or less, directly or indirectly, it by no means follows that it is everybody's particular responsibility. By established rule and justice, the determination belongs primarily to those immediately on the spot, in actual possession. Unhappily, the powers that border the Persian Gulf, Persia itself, Turkey, and some minor Arabian communities, are unable to give either the commercial or the military security that the situation will require. Under their tutelage alone, without stronger foundations underlying, stability cannot be maintained, either by equilibrium or by predominance. In such circumstances, and when occasion arises, the responsibility naturally devolves, as for other derelicts of fortune, upon the next of kin, the nearest in place or interest. If they, too, fail, then the more remotely concerned derive both claim and duty. The general welfare of the world, as that of particular communities, will be most surely advanced by each one doing that which he finds to his hand to do, whether by direct charge received from due authority, or by inheritance, or from the mere fact of neighborhood, which has given to the word "neighbor" that consecrated association, with the sound of which we are all familiar, though we too narrowly conceive the range of its privilege and its duty.

From the fact of propinquity, of geographical nearness, or of direct political interest, it is easy to see that Great Britain and Russia are the two States which from existing circumstances are most immediately and deeply concerned; nor, when the several circumstances are closely analyzed and duly weighed, does there to my mind seem room to doubt that to the former falls first to say whether she will discharge the duty, or let it go to another. Let there be here interposed, however, the word of caution, before quoted, concerning the natural course of things, lest I should seem fairly chargeable with the disposition, unwise as well as unjust, to favor needless or premature intervention. It may well to-day be a duty not to do that which to-morrow will find incumbent. Opportunity is not to be created, but to be awaited till it appear in the form of necessity, or at the least of clear and justifiable expediency. Consulting the natural order of things, forcing nothing, means at least invincible patience as well as sleepless vigilance; and vigilance includes necessarily readiness, for he only is truly awake who is careful to prepare.

I have said that an analysis of the circumstances shows that Great Britain, in the evident failure of Turkey and Persia, is the nation first—that is, most—concerned. She is so not only in her own right and that of her own people, but in the yet more binding one of imperial obligation to a great and politically helpless ward of the Empire; to India and its teeming population. In her own right and duty she is, as regards the establishment and maintenance of order, in actual possession, having discharged this office to the Gulf for several generations. Doubtless, here as in Egypt, now that the constructive work has been done, she might find others who would willingly relieve her of the burden of maintenance; but as regards such transfer, the decision of acceptance would rest, by general custom, with the present possessor. To her the question is one not merely of convenience, but of duty, arising from and closely involved with existing conditions, which are the more imperative because they are plants of mature growth, with roots deep struck and closely intertwined in the soil of a past history.

These conditions are doubtless manifold, but in last analysis they are substantially three. First, her security in India, which would be materially affected by an adverse change in political control of the Gulf; secondly, the safety of the great sea route, commercial and military, to India and the farther East, on which British shipping is still actually the chief traveller, though with a notable comparative diminution that demands national attention; and, thirdly, the economic and commercial welfare of India, which can act politically only through the Empire, a dependence which greatly enhances obligation. The control of the Persian Gulf by a foreign State of considerable naval potentiality, a "fleet in being" there, based upon a strong military port, would reproduce the relations of Cadiz, Gibraltar, and Malta to the Mediterranean. It would flank all the routes to the farther East, to India, and to Australia, the last two actually internal to the Empire, regarded as a political system; and although at present Great Britain unquestionably could check such a fleet, so placed, by a division of her own, it might well require a detachment large enough to affect seriously the general strength of her naval position. On the other hand, India, considered in regard to her particular necessities, apart from the general interests of the Empire, may justly demand that there be secured to her untrammelled intercourse with Mesopotamia and Persia. She has a fair claim also to any incidental advantage attendant upon the through land communication that can be assured by political foresight, obtaining a position favorable to the negotiations of the future. It is notorious, for instance, that most nations, and Russia pre-eminently, adopt a highly protective or exclusive policy towards foreign industries. Applied to what is now Persia, this would be a direct injury to India, which, even under the present backward conditions of the inhabitants and of communications, carries on a large part of the Persian trade, as might naturally be expected from the nearness of the two countries. The same is doubtless true of her relations with Mesopotamia, though the absence of reliable customs returns prevents positive statements. For securing these natural rights of India, British naval predominance in the Gulf, unfettered by bases there belonging to possibly hostile foreign powers, would be a political factor of considerable influence; but it is incompatible with the establishment of foreign arsenals.

Further, purely naval control is for this purpose a very imperfect instrument, unless supported and reinforced by the shores on which it acts. It is necessary therefore to attach the inhabitants to the same interests by the extension and consolidation of commercial relations, the promotion of which consequently should be the aim of the government. The acquisition of territory is one thing, which may properly be rejected as probably inexpedient; and certainly unjust when not imperative. It is quite another matter to secure popular confidence and support by mutual usefulness. Whatever the merits of free trade as a system, suited to these or those national circumstances, it probably carries with it a defect of its qualities in inducing too great apathy towards the exertion of governmental action in trade matters. Non-interference, laissez-faire, may easily degenerate from a conservative principle to an indolent attitude of mind, and then it is politically vicious. The universal existence and the nature of a consular service testify to the close relationship between trade and government, a relationship that is in some measure at least one of mutual dependence. A certain forecast of the future, a preparation of the way by smoothing of obstacles, a discernment of opportunity,—which is quite different from creating it,—a recognition of the natural course of things at the instant when it may be taken at the flood, these are natural functions of a competent consular body. To it belongs also the establishment of international relations through the medium of personal intercourse, so strongly operative in public matters even in states of European civilization, among statesmen whose business is to look below the surface, and beyond the individual, to the substantial and permanent issues at stake. Much more is it influential among peoples where statesmanship is chiefly a matter of personal interest or bias, consequently short sighted and unstable, and where local confidence and prestige are dominant factors in sustaining policy. There the flag, if illustrated in a well-organized consular service, may well be the forerunner of trade as well as its necessary complement.

At the present time the trade of Persia is divided chiefly between Great Britain and India on the one hand, and Russia on the other. As would be expected from their relative positions, the northern part falls to Russia, the southern to her principal rival in Asia. The one therefore is essentially a land trade, the other maritime. From these respective characteristics, the one naturally induces governmental intervention, to promote the facility of communications, to which the land by its varied and refractory surface presents continual obstacles. The other finds its royal highway of the sea ever clear and open, a condition which ministers to the natural conservatism and acquired principle of non-interference which distinguish Great Britain. By the disposition of all living things to grow, the spheres of the two tend continually to approach. The moment of contact may well be indefinitely distant, but the circumstances which shall attend its arrival are already forming; and when it comes it may be, as now in China, the signal of an antagonism, the result of which will depend upon the facts of political position on the one side or the other. Russia not unnaturally looks to her continuous territory and population, behind the scene of possible contest, as the assurance of her own permanent predominance and eventual exclusive influence. It may be so; but not necessarily until a future so far distant as to be utterly beyond the range of our possible vision, and between which and us lie many chapters of unknowable changes. If confronted by a solid political organism, resting immediately upon commercial interests, and ultimately upon naval control of the Gulf and the armed forces of Great Britain, backed by her colonies and India, it must be long before the northern impulse can overcome the resistance. The physical difficulties of the land route contrasted with the level path of the sea, the narrowness of rail carriage as compared with the broad highway of the ocean, more than compensate for the apparent shorter distance and delusive continuity of the land. The energies of Russia also must long be absorbed by other necessary pre-occupations, notably the far superior importance of developed and consolidated access, by Siberia and Manchuria, to North China seas and the Pacific, the great immediate centres of world interest. There is therefore no need to hasten things in their natural course, but equally there is no justification for neglecting to note and improve them; to quote Washington again, "diffusing and diversifying by gentle means the streams of commerce," which will gradually nurse the future into vigorous life.

Both Persia and China are being swept irresistibly into the general movement of the world, from which they have so long stood apart. Both have a momentous future of uncertain issue, but that of China is evidently more immediately imminent. This is the natural course which things are at present following. Persia has still a time of waiting. The indications also are that Russia, consciously or intuitively, thus reads the conditions. By farsighted sagacity, or through continued yieldings to the successive leadings of the moment, she has now extended her great effort towards sustained communication with ever-open water to the farther East. The Siberian railroad, by which she hopes to assure it, passes through territory that is wholly her own by ancient tenure; while through recent generations she has prepared its security by her steady progress southward in Central Asia and Turkestan. The establishment of orderly government in those regions relieves the flank of the route from predatory dangers, which under the feeble administration of Turkey will constitute one of the elements of difficulty for the projected railroad in the Euphrates valley. The Siberian road throughout its whole course is unassailable by any external power, until within a very short distance of the coast terminus. Its military safety being thus absolute, its maintenance, and the development of its carrying power, essential to the Russian position in the farther East, are questions simply of money. Money, however, will be needed in such quantities that the imperative requirements must postpone further effective movement to the southward or westward; for effective movement means developed communications, consolidated and sustained. These are expensive, and in sound policy should not be attempted on a grand scale in two directions at the same time; unless indeed the resources in money and labor are so great as to justify their dissemination. That this is not the case the notorious condition of the Siberian road gives reason to believe.

Water communication with the external world, through an unimpeded seaboard of her own, is Russia's greatest present want. For this object, to what extent would she benefit commercially by access to the Persian Gulf, as compared with the China seas? Putting out of consideration China itself, with the nearer shores of the Pacific, as to which the better situation of Manchuria cannot be questioned, Russia is there much closer also to the Americas and to the entire Pacific. Australia is substantially equidistant from the Persian Gulf and from Port Arthur; the balance favoring the latter. Only Southern Asia and Africa can be said to be nearer to the Gulf. Europe and Atlantic America are now reached, and ever must be reached, commercially, by Russia, from the Black Sea or the Baltic. From the standpoint of military advantage, a Russian naval division in the Persian Gulf, although unquestionably a menace to the trade route from Suez to the East, would be most excentrically placed as regards all Russia's greatest interests. It is for these reasons that I have elsewhere said that the good of Russia presents no motive for Great Britain to concede a position so extremely injurious to herself and her dependencies.

The question of the Persian Gulf, and of South Persia in connection with it, though not yet immediately urgent, is clearly visible upon the horizon of the distant future. It becomes, therefore, and in so far, a matter for present reflection, the guiding principle of which should be its relation to India, and to the farther East. This again is governed by the strategic consideration already presented in the remark that movement, advance, to be effective and sustained, requires communications to be coherent and consolidated. The Russian communication by land, though still inadequately developed, is thus secure, militarily. Throughout its length there exists no near-by point held by an enemy able to interrupt it by a serious blow. The significance of such a condition will be realized forcibly by contrasting it with the military exposure of another great transcontinental line, the Canadian Pacific. In the farther East Great Britain, like Russia, holds an advanced position, chiefly commercial, but consequently military also, the communications of which are by water. These have not, and probably never can have, any military security comparable to that of the Siberian railway. Their safety must depend upon sustained exertion of mobile force, resting upon secure bases, ready for instant and constant action. It is needless to insist upon the difficulty of such a situation; it has been made the subject of recent and abundant comment. But if thus onerous now, all the more reason that the burden should not be increased by the gratuitous step of consenting, upon any terms of treaty, any forced infringement of the natural condition of things, to the establishment of a new source of danger analogous to those already existing in Cadiz, Toulon, the Dardanelles, and so on. Concession in the Persian Gulf, whether by positive formal arrangement, or by simple neglect of the local commercial interests which now underlie political and military control, will imperil Great Britain's naval situation in the farther East, her political position in India, her commercial interests in both, and the imperial tie between herself and Australasia.

So far from yielding here, it appears to me that the signs of the times, as outlined above, point seriously to the advisability of concentrating attention, preparation of the understanding at least, upon that portion of the Suez route to the farther East which lies between Aden and Singapore. In this the Persian Gulf is a very prominent consideration. It is not necessary that material preparation should far forestall imminent necessity; but the preparation of thought which we call recognition, and appreciation, costs the Treasury nothing, and saves it much by the quiet anticipation of contingencies, and provision against them. It tends to prevent inopportune concessions, and the negligences which arise from ignorance of facts, or failure to comprehend their relations to one another. The South African War and the twenty preceding years give recent warning. Foreign affairs, as well as military, need their general staff. Besides its bearing upon the Suez route, the Gulf has a very special relation to the Euphrates valley, and any road passing through it from the Levant; and this relation is shared by South Persia, because of the political effect of its tenure upon the control of the Gulf. There is here concentrated therefore commercial and political influence upon both of the two routes, that by land and that by water, from the Mediterranean to India and to the East beyond. There is no occasion in the nature of things that Great Britain, either by concession or compulsion, should share with another State the control which she now has here; but in order to retain it she needs not only to keep the particular protective relations already established with minor local rulers, but further to develop and fortify her commercial interests and political prestige in South Persia and adjacent Mesopotamia. This means not only, nor chiefly, increase of exchange of products. It means also partnership, public or private, in the system of communications, analogous in idea, and if need be even in extent, to Disraeli's purchase of the Suez canal shares. The attitude of the United States Government towards the projected Panama Canal affords a further suggestive illustration. As towards the farther East, South Persia is in fact the logical next step beyond Egypt; though it does not follow that the connection therewith is to be the same. Correlative to this commercial and political progress, goes the necessity of local provision for naval activity when required. The middle East, if I may adopt a term which I have not seen, will some day need its Malta, as well as its Gibraltar; it does not follow that either will be in the Gulf. Naval force has the quality of mobility which carries with it the privilege of temporary absences; but it needs to find on every scene of operation established bases of refit, of supply, and, in case of disaster, of security. The British Navy should have the facility to concentrate in force, if occasion arise, about Aden, India, and the Gulf.

In summary: Relatively to Europe the farther East is an advanced post of international activities, of very great and immediate importance; but from the military point of view, to which as yet commercial security has to be referred, the question of communications, of the routes of travel, underlies all others and must be kept carefully and predominantly in mind. Russia has her own road, by land, unshared with any other. To the rest of Europe, and to Russia when she chooses, there exists now the sea route by Suez, which is, and probably must remain, supreme to all others. Alternative to it, in part of the way, the future will doubtless bring railways. These, however, on account of the greater cheapness of water carriage, will pretty surely do their principal through business in expediting special transit between the two seas—the Mediterranean and the Indian Ocean. They will in this respect maintain merely an express and fast freight traffic. Between them and the Suez route there will be the perennial conflict between land and water transport, between natural and artificial conditions, in which the victory is likely to rest, as heretofore, with nature's own highway, the sea. But, however that prove, the beginning and the end, the termini, of both routes, land and sea, so far as they compete, will be substantially the same: the Levant Sea, the Straits of Bab-el-Mandeb and the Persian Gulf. It is too much to ask of international compliancy that Europe should accept the single control of both terminal regions by the same State, especially where no defined claim now exists, as is the case in Levantine Turkey; but equally, where a single government can show a long prescription of useful action, of predominant influence, and of political primacy locally recognized in important quarters, as Great Britain can, there is no reason why she should be expected to abandon these advantages, except as the result of war, if a rival think that result will repay the cost.

There is not to be seen in the nature of things any evidence, or any tendency, which indicates the probability that Great Britain may be forced to yield to compulsion, actual or threatened, concessions of present right which it is inexpedient that she should grant voluntarily. It is upon such probability, conceived to be imminent, that are based proposals of arrangement, or compromise, that I cannot but think excessively artificial, and disregardful of permanent conditions. They surmise, as a necessary postulate, hostile combinations of two or more States, against which, by a curious intellectual prepossession, no probable counterpoise is discernible. As a matter of fact, founded upon present territorial positions, there is in the nature of things no real, no enduring, antagonism concerning the Persian Gulf, except between Great Britain and Russia. It is not to the interest of any third State to interfere between these two, or to disturb—much less to destroy—the local balance of power which now exists between them and can probably be maintained. As regards its particular interests, the hands of any third State will be not more, but less, free, should that balance yield to the decisive predominance of one of the two throughout the regions involved. Nor can a third State expect to restore equilibrium, if lost, by itself taking the place of the one that has gone under. It is only necessary to consider the solidity, extent, and long standing, of the local control now wielded by Russia and Great Britain, together with the land power of the one and the sea power of the other, to see the hopelessness of any substitute for either in its own sphere. The two systems are not dead, but living; not machines, but organisms; not merely founded, but rooted, in past history and present conditions. What the rest of the world needs, what world politics requires, is that here, as in Asia immediately to the eastward, there should be political and military equipoise, not predominance. The interests of other States are economical; freedom of transit and of traffic, the open door. The very problem now troubling nations in the Levant and China is how to establish,—and only afterwards to maintain,—conditions which are already established and have now only to be maintained about the land approaches to the Persian Gulf.

There is therefore, no sound inducement for another State to waste strength here. It can be used better elsewhere. When substantial equilibrium thus exists, a slight effort will suffice to obtain from either party a consideration which in the case of distinct predominance, or exclusive tenure, might require a full display of national power. Doubtless, many in Great Britain, and also in America, are convinced that one third State, the German Empire, is restlessly intent, not only upon economical and maritime development, which is not to be contested by other than economical weapons, but also upon self-assertive aggression with a view to territorial aggrandizement in more than one part of the world; and notably in this particular quarter. A concession has been granted to German capitalists to extend the railway, which now ends at Konieh, to Bagdad, passing through the Euphrates valley. The necessary outlet to this is the Persian Gulf. Such concession, when realized in construction, carries with it a national investment, an economical interest, which, though in private ownership, inevitably entails political interest. It justifies public backing by its own government, in countries where, as in Turkey, private right is secure only when it has national force behind it. It is for this very reason that Great Britain, having already political interest in the Persian Gulf, should encourage British capital to develop communications thence with the interior in Persia and in Mesopotamia, as strengthening her political claim to consideration, and excluding that of possible antagonists. The German road would thus find its terminus in a British system; a not unusual international relation. German enterprise has in anticipation established German political hold upon Asia Minor and Mesopotamia. As expectation passes into realization Germany will acquire local political importance and influence; a right, sanctioned by the rules of intercourse with Oriental nations, to have her voice heard in many local matters, as affecting the interests of her subjects who are thus engaged in developing the country.

What effect will this have upon Germany's political and military position, relatively to Russia and Great Britain, which, from nearness or from the commercial ubiquity of their citizens, are also politically interested? Under present conditions Germany, whose nearest port is in the North Sea, has assumed a political burden at a point from which she is far more remote than Russia, and her sea approach to which is before the face of the much greater navy of Great Britain. There is in this nothing to prevent the just assertion of her right, no necessary cause of quarrel,—far from it; but also there is nothing menacing. Germany has simply introduced another factor into a problem as yet unsolved, that of the ultimate political status of several provinces of the Turkish Empire,—Asia Minor, Syria, and Mesopotamia. As I have elsewhere said, I believe that her appearance there is a step towards a right final solution; that from the necessary common interest of Germany and Great Britain in the Suez route to the farther East, because the commerce of both depends upon its security, the two cannot but work together to secure here a political development which will consolidate their respective naval positions in the Levant.

This seems to me an absolute permanent condition, consistent with a certain amount of mutual jealousy and political wrangling, and with unlimited commercial rivalry, but nevertheless determinative of substantial co-operation. The mass of Russia is so vast, her ambitions so pronounced, and she is so near at hand, that the Suez route needs precisely that kind of protection against her which Russia herself has given to the Siberian road by the regularization of the provinces south of it. Whatever the particular form local administration may ultimately assume, it is imperative upon the Teutonic States to see that their water route to the East is not imperilled by naval stations flanking it, whether in the Levant or in the Persian Gulf. Being themselves far distant, dependent upon naval power simply, it is essential that they constitute a political pre-occupation favorable to themselves in the Asian provinces of Turkey and in Southern Persia. In Egypt and in Aden Great Britain has already done much. Germany, in building a Mesopotamian railway, the continuation of that already working in Asia Minor, contributes to the same end. That Russia looks upon the enterprise with disfavor is a testimony, conscious or unconscious, to its tendency.

These also seem to me permanent considerations. Not less so, having reference to the anxiety felt by some in Great Britain as to the intentions of Germany, is the general situation of the latter in European politics. There is certainly an impression in America, which I share, that Great Britain for various reasons has been tending to lose ground in economical and commercial matters. Whether this be a passing phase, or a symptom of more serious trouble, time must show. Should it prove permanent, and Germany at the same time gain upon her continuously, as for some years past she has been doing, the relative positions of the two as sea powers may be seriously modified. The danger appears to exist; and if so the watchmen of the British press should cry aloud and spare not until all classes of their community realize it in its fundamental significance. Military precautions, and the conditions upon which they rest, have been the main motive of this paper; but these, while they have their own great and peremptory importance, cannot in our day, from the point of view of instructed statesmanship, office-holding or other, be considered as primary. War has ceased to be the natural, or even normal, condition of nations, and military considerations are simply accessory and subordinate to the other greater interests, economical and commercial, which they assure and so subserve. In this article itself, turning as it does on military discussion, the starting point and foundation is the necessity to secure commerce, by political measures conducive to military, or naval, strength. This order is that of actual relative importance to the nation of the three elements—commercial, political, military.

It is evident, however, that these primary matters, although they underlie this argument, are otherwise outside it. For the rest, as regards the general military strength, and in particular the sea power, of the two countries, nothing can overthrow the one permanent advantage that Great Britain enjoys in being insular. Germany, should she realize her utmost ambitions, even expanding to the Mediterranean, must remain a continental State, in immediate contact with powerful rivals. Historically, no nation hitherto has been able under such conditions to establish a supreme sea power. Of this France is the historical example. On the other hand, regarded in herself alone, apart from rivals, Germany cannot, as the United States could not, exert the intense internal effort now required for political consolidation and economical development coincidently with an equal expansive effort. The one may succeed the other, as in our case and in that of Great Britain, where the expansion of the eighteenth century followed and depended on the unifying action of the seventeenth; but, until internal coherence is secured, external expansion cannot adequately progress. One weakens the other. Though correlative, they are not co-operative.

The ambition of Germany so to develop her fleet as to secure commercial transit of the North Sea, which washes her entire maritime frontier, is a national aspiration in itself deserving of entire sympathy. Towards all other States except Great Britain it is within the compass of reasonable expectation. As towards Great Britain it is, under present economical conditions, impossible; for Great Britain, being insular, must maintain continuously supreme the navy upon which her all depends, and moreover, as I pointed out in a recent paper, by geographical position she lies across and flanks every sea route by which Germany reaches the outer world. This condition is permanent, removable only by the friendship or destruction of the British power. Of the two the friendship will be the cheaper and more efficacious; for it is needed not in home waters only but in those distant regions which we have been considering. The naval power of Great Britain is just as real a factor in the future of Germany in the distant East as every thinking American must recognize it to be in our own external policy. That such a force should be paid for, and must necessarily be maintained, by another people, whose every interest will prompt them to use it in the general lines of our own advantage, is a political consideration as valuable as it is essentially permanent. In the matter of exertion of force it accords absolutely with the nature of things. As for economical rivalry, let it be confined to its own methods, eschewing force.

In saying these things I may seem to ignore the bitter temper, openly and even outrageously shown by the German people towards Great Britain in these last three years. I do not forget it. Human nature being what it is, the dangerous effect of such conditions upon international relations is undeniable. It is ever present to my reflections upon the political future. The exhibition is utterly deplorable, for it can serve no good end, and if it continue will prevent a co-operation among the three Teutonic States which all need, but Germany most of all; for the respective external interests of the United States and Great Britain—together with Japan—have so much in common, and so little that is antagonistic, that substantial, though informal, co-operation is inevitable.

This hostility constitutes an element in the political situation which should be taken into account, and carefully watched. Nevertheless, the permanent conditions, above summarized, will through a future beyond our possible present foresight retain Germany in a position of naval numerical inferiority to Great Britain, as regards both mobile force and the essential naval stations which the latter has acquired during two centuries of maritime activity. These conditions, by their inevitable logic, ought ultimately to overcome a sentiment which has no good ground for existence, and which betrays the national interest. Should it, however, endure, the permanent facts are too strong for it to do more than dash harmlessly against them. Awaiting either event, may not the people of Great Britain on their part, without relaxing vigilance or ignoring truths, accept Washington's warning, which we Americans at least have by no means outgrown, against "permanent inveterate antipathies against particular nations." They have cause for anger; but anger disturbs the judgment, and I think in some measure is doing so in this instance. This particular antipathy is yet young, let it not harden into maturity. In the great political questions which for some time to come will concentrate the external regard of nations and statesmen, the natural desires of Russia, reasonable and unreasonable, are contrary to those of Germany as well as of Great Britain. It is to her clear interest that they remain alienated. Such conditions should on the one hand prompt an earnest effort for a balanced and conciliatory adjustment on all sides; but on the other, their essential permanence, if it be as I think, demands a recognition which would show itself in the extrusion of everything resembling passion, and in the settlement of national purpose on the firm ground of essential facts, instead of the uncertain foundation of any artificial agreement which contravenes them.

CHAPTER X

THE NAVAL WAR COLLEGE

IN military activities the question of the utilization of the armed forces is the most critical and the most vital that confronts a nation. Utilization in war is the final stage of a progress which begins with the drill-ground, where the raw recruit is fashioned into the finished soldier, and with the workshops where crude material is converted into weapons of war. Utilization presupposes all the successive processes of organization and equipment; whereby, step by step, out of individual men are built up huge military units, army and army corps, battle-fleet and battle-ship, as individual in their power of intelligent corporate action as is the one man in his single existence. Thus, assuming the foundations upon which action rests, the directing authority dismisses them out of mind, concentrating attention purely upon the problem how best *to use* those entities which organization and equipment have supplied. It is to a similar concentration I would here invite readers, asking them also to dismiss from their minds, as not under consideration, all thought of the material of war, of the antecedent processes by which a national fleet or national army is built up; to accept each and both as being ready, with only the one question remaining: how they, or either of them, is *to be used* to the best advantage in war?

North American Review, vol. 196 (July 1912), 72–84. Reprinted in *Armaments and Arbitration* (Boston, 1912), 196–217.

The methods by which this result is to be reached are divided naturally under three heads. These, in the order of time sequence, are Movement, Strategy, and Tactics. The first of these comprises not only motion, but all the dispositions for marches and transportation of supplies which make possible the transference of armies over ground, in advance or in retreat. This function of moving armies and their trains has received the technical name Logistics. Various derivations have been assigned for this term; the one now generally accepted is from a Greek word, the root idea of which is "calculation." It is not necessary to enlarge upon the complications of detail involved in moving huge bodies of men, with their supply-trains, by calculated progress, stage by stage; including each day's march, each day's halt, each day's meals, over roads in any case relatively narrow. All this may be assumed, or left to the imagination. But it should be observed that the special characteristic of this class of operations is movement, pure and simple. The movement, it is true, is minutely organized in many intricate particulars, and therefore is truly a work of military art; but withal it is not accompanied by those particular directive ideas which in strategy and tactics make movement subordinate to action, in which movement is in itself merely contributory. In short, in logistics movement is the principal; whereas in strategy and tactics it is only an agent.

In sea warfare the analogue of logistics is found, but much simplified in conception by that quality which is the distinguishing characteristic of sea forces—mobility. Mobility facilitates supply, as it does the movement of the fleet itself. The narrow strip of marching surface afforded even by the greatest highways is superseded by the broad bosom of the deep. The ocean presents no natural impediments, few obstacles. Each ship carries stores for weeks; and at night there is no halt, no wait for food-supplies. The vessels move straight on for their goal with unwearied crews. The necessary train of supply-ships, repair-vessels, colliers, all have mobility like to that of the fleet itself. But there remains a counterbalancing factor affecting the question of sea logistics: that of sustained movement and maintenance during a campaign. Fleets more often than not operate remote from home. Consequently, the chief items of supply must traverse long sea distances, under conditions of exposure exceeding the corresponding chain of supplies of an army, which in their approach are secured in large measure by the interposition of the army itself between them and the enemy; a safeguard expressively phrased in the words "covering the communications." In such case land communications may suffer by a raid, unexpectedly and momentarily; but raids by land are restricted in time and space by the imperfect mobility inherent in land conditions, whereas the mobility which is the prerogative of the water makes sea communications much more liable to successful harassment.

It will be recognized, therefore, that the determining the places of rendezvous for coal and other supplies, the protection of the routes, the whole question of keeping the holds and coal-bunkers full, and the several ships in best steaming condition, is a big administrative calculation and co-ordination, which is an instance of logistics because it directly affects the fleet's power of action. Nelson, by diligent watchfulness, always during his last great campaign had his ships stored full for three months; usually for five. That is, the movement of his fleet, wherever he would, was assured for those periods. Wrote a contemporary to him:

> You have extended the powers of human action. After an unremitting cruise of two long years in the Gulf of Lyons, to have proceeded, *without going into port,* to Alexandria [in Egypt], from Alexandria to the West Indies, from the West Indies back again to Gibraltar; to have kept your ships afloat, your rigging standing, and your crews in health and spirits, is

> an effort such as never was realized in former times. You have protected us for two long years, and you saved the West Indies by only a few days.

This was an achievement of logistics, of movement constant and unimpaired, because of diligent prevision. No fighting; yet it underlay Trafalgar.

Yet it is very different from the battle Trafalgar, which illustrates tactics; different also from the various movements of the British and hostile fleets in the half-year before Trafalgar, in which there was abundance of motion directed toward specific points and with specific aims, covering both the Atlantic and the Mediterranean. The general conceptions underlying such specific aims are known as strategy; the movement of ships in furthering them was merely a contributory agent, which resulted in bringing the fleet to the scene of action. In like manner the movement of the ships in the battle was merely contributory, to carry out the tactical conception of the method of attack.

From the outline sketch of logistics here presented it is evident that it is an immense administrative function, covering many details and requiring much system and prevision, justifying the derivation from "calculation." In management, however, it is somewhat deliberate, and should fall mainly upon men subordinate in office to those who guide the great military conceptions of strategy and tactics. Logistics is dwelt upon first because, while as vital to military success as daily food is to daily work, yet, like food, it is not the work. In this paper attention is to fasten upon the work. Like organization and equipment, logistics underlies achievement; but while nearer the field of battle than those are, and coincident and contemporary with the action of the field, logistics yet is not, so to say, on the fighting-line, nor has it to do with the direction of those movements upon which success and victory immediately hinge.

Evidently the management of such a system of movement

and supply requires much experience, and also that training or instruction which in most professions precedes experience and facilitates its acquisition. Similarly, training and experience are requisite in the more advanced stages of the art of war; in strategy and tactics. And it is to be noted closely, as well as clearly, that the object of training and instruction is not merely to mold the individual, but to impress upon each a common type, not of action only, but of the mental and moral processes which determine action; so that within a pretty wide range there will be in a school of officers a certain homogeneousness of intellectual equipment and conviction which will tend to cause likeness of impulse and of conduct under any set of given conditions. The formation of a similar habit of thought, and of assurance as to the right thing to do under particular circumstances, reinforces strongly the power of co-operation, which is the essential factor in military operations. Combination and concentration, two leading ideas and objects in war, both indicate unity of energy produced by the harmonious working together—co-operation—of many parts.

Obviously such harmony is not best when merely mechanical, for a mechanical mind is easily deranged in presence of the unexpected. It is the inspiration of common purpose and common understanding which, when the unexpected occurs, supplies the guiding thought to meet the new conditions and bend them to the common end. If this condition be adequately attained, the mind of the commander-in-chief will be omnipresent throughout his command; the most unexpected circumstances will be dealt with by his subordinates in his spirit as surely as though he were present bodily. It is difficult to overestimate the importance of such a result. The captains of individual battle-ships, the commanders of the several corps of an army, have it in them to make or to mar the purposes of the commander-in-chief; not by disaffection, but by lack of comprehension. Lord Howe's entire plan of battle in

1794 was thus wrecked, as was Rodney's on an earlier occasion, by incapacity which previous training should have obviated. In land warfare the twin battles of Gravelotte and St. Privat, in the Franco-German War, gave illustration, one of a subordinate fully comprehending, and consequently not only executing his general's full conception, but developing it even further as opportunity arose; whereas the other, by failure to comprehend, effected merely confusion and disorder, without result.

It is to supply such common understanding and inspiration that war colleges have been instituted. Those who receive the training go forth imbued with a common mode of thought, which latterly has received the name of "Doctrine." There is about this word a suggestion of pedantry which impels to a justification for the use of it. In military operations doctrine, if not given the name, has always existed. When Nelson took his first independent command, three months before the battle of the Nile, he summoned his captains frequently on board his own ship, where he explained to them his proposed methods of action under many possible conditions. This was his doctrine. When the battle came off, each captain understood what he was to do and what the others were to do; and not mechanically, but with a general idea applicable to all probable circumstances. "I should never have dared to attack as I did without knowing the men, but I was sure each would find a hole to creep in at." Each captain was possessed with the spirit and understanding of Nelson himself.

In like manner before Trafalgar, the *Nelson touch* of which he spoke exultingly was the Nelson doctrine, imparted to the captains severally and collectively, and by them received enthusiastically. "It was my pleasure to find it not only generally approved, but clearly perceived and understood." Collingwood's impatient remark when Nelson made his famous last signal, "I wish Nelson would stop signaling, for we all know what we have to do," is an affirmation of "doctrine" understood. An imperfect comprehension of Rodney's doctrine by the captain whose ship was the pivot of the operation lost the admiral what he considered the greatest opportunity of his life. The absence of "doctrine" is shown by his words subsequently:

> I gave public notice that I expected implicit obedience to every signal made. My eye on them had more dread than the enemy's fire, and they knew it would be fatal. In spite of themselves I taught them to be what they had never been before—officers.

It is to be observed that the eye of the admiral had to be everywhere, just because there was among the officers no spirit of doctrine on which he could rely.

The French word *doctrinaire,* fully adopted into English, gives warning of the danger that attends doctrine; a danger to which all useful conceptions are liable. The danger is that of exaggerating the letter above the spirit, of becoming mechanical instead of discriminating. This danger inheres especially in—indeed, is inseparable from—the attempt to multiply definition and to exaggerate precision; the attempt to make a subordinate a machine working on fixed lines, instead of an intelligent agent, imbued with principles of action, understanding the general character not only of his own movement, but of the whole operation of which he forms part; capable, therefore, of modifying action correctly to suit circumstances. "When I tell Lord Howe to do anything," wrote his senior, "he never asks how it is to be done, but goes and does it." This illustrates the proper relation of a superior to a subordinate. It is not only generous, but sagacious. Hence, in the instruction of war colleges great stress is laid upon the formulation of orders; in the particular respect that while they are to convey lucidly to the subordinate the general aim of the operation, and his own specific share, with such collateral factors as are necessary for his understanding of the situation, the guidance is left in his hands. He is to be told what is to be done, not hampered with directions how to do it; because the "how" may not fit a condition he finds before him, but even more because his own power of independent initiative is too valuable a military asset to be so repressed.

A curious illustration of the existence of a doctrine, among seamen not usually suspected of theorizing but considered specifically practical, is found two hundred years ago in the express order of the British Fighting Instructions that an attacking fleet was first to form on a line parallel with the enemy and then to steer down upon him, all ships together; the van to engage the enemy's van, the center the center, the rear the rear. It was a very bad doctrine; not least bad in that it took all discretion away from every one. The one saving clause—unexpressed—was that a man who fights will always be approved. Contrast this with Nelson at St. Vincent. It is true he had received no doctrine from his commander-in-chief, but he had the equivalent—he perceived his senior's plan; and, seeing it about to fail, he broke out of the order and thwarted the enemy's attempt. Brilliant as this was in an exceptional man, it is much better that the average man should be equipped with the understanding which would reach the same result through comprehension. Collingwood, a distinguished example of the average man, was on this occasion close behind Nelson, in the order most favorable situated to imitate him; but he had no doctrine by which to overpass the signals.

It seems self-evident that if a doctrine, as described, is to be valid to the ends of a common spirit and to foster individual power of initiative on certain broad common lines, it must be not only a general principle, or set of such principles, but must be assimilated mentally through numerous illustrations. In other words, it must be based on antecedent experience. Formulated principles, however excellent, are by themselves too abstract to sustain convinced allegiance; the reasons for them, as manifested in concrete cases, are an imperative part of the process through which they really enter the mind and possess the will. On this account the study of military history lies at the foundation of all sound military conclusions and practice. It therefore is the basis, the corner-stone, upon which the instruction of a war college rests. Historical occurrences, analyzed and critically studied, have been the curriculum through which great captains have trained their natural capacity for supreme command. They correspond to the legal cases and precedents which embody and illustrate principles, and so govern legal argument and judicial decision, the struggle and the victory of a court of law.

It is evident on consideration that military precedents derived from history are chiefly valuable as embodying principles, which are to be elicited and then to be applied in circumstances often very different. They are not mere models for a copyist. Two battles will rarely be fought on the same ground; and were the ground the same, the constitution and numbers of the opposing forces will vary. A leading feature in war-college instruction, therefore, necessarily is the constitution of new cases, problems, hypothetical but probable, to the solution of which are to be applied the principles derived from military history. The applicatory system, as it is fitly called, is thus the superstructure, raised upon the basis of experience as embodied in historical military events. It is to be observed that this system, though artificial, reproduces closely the conditions under which military decisions have to be reached in actual war. Each situation that arises in the course of a campaign is a new case, to which the commander-in-chief applies considerations derived from his own experience or from his knowledge of history. It is not meant that these applicatory processes in the field are always conscious efforts of memory, although Napoleon has said that on the field of battle the happiest inspiration is often only a recollection. The exercise of the functions of a trained mind is instinctive, as well in such recollection as Napoleon cited as in decisions which seem wholly personal. Said the great Austrian general, the Archduke Charles:

> A general often does not know the circumstances upon which he has to decide until the moment in which it is necessary to proceed at once with the execution of the necessary measures. Then he is forced to judge, to decide, to act with such rapidity that it is indispensable to have the habit of embracing these three operations in a single glance. But that piercing perception which takes in everything at a glance is given only to him who by deep study has sounded the nature of war; who has, so to speak, identified himself with the science.

This is a tribute to the methodical training of faculties. Such training is the peculiar object of the applicatory system—to identify the mind and its habit of action with the art of war, by continuous exercise in dealing with numerous varied instances; a process of repetition which cannot but have the effect that habit always has upon conduct and character. The statement of this effect appeals to the experience of every one. All know how inevitably and unconsciously one repeats the same action under similar circumstances—the "second nature" of the proverb. When this result has been produced in a number of men who act together, there will extend throughout the entire command a unity of purpose and of comprehension which to the utmost possible extent will insure co-operation, because it has already insured a common understanding and habit of action. Thus of the renowned Light Brigade of the Peninsular War, formed under the still more renowned Sir John Moore, it is said that "the secret of its efficiency lay in inculcating correct habits of command in the regimental officers."

> The system of discipline, of instruction, and of command formed in the persons of their company officers a body of intelligent and zealous assistants, capable of carrying out the

> plans and *anticipating* the wishes of their seniors; not merely a body of docile subordinates capable of obeying orders in the letter, but untrained to resolute initiative. The most marked characteristics of Sir John Moore's officers were that when left alone they almost invariably did right. They had no hesitation in assuming responsibility. They could handle their regiments and companies, if necessary, as independent units; and they consistently *applied* the great *principle* of mutual support.

A convenient, because recent, instance of an actual case, which might very well have been constituted as suppositive by an instructor, may be found in the circumstances and conditions of the respective military and naval forces of Japan and Russia before their still recent war. The Japanese authorities had before them the positions of the Russian principal army in Manchuria, the fortified port of Port Arthur, the actual or estimated numbers in the field and in the garrison, the Russian main fleet in Port Arthur, the powerful detachment in Vladivostok, the Russian vessels on the way east at various points; probably also the two or three at Chemulpo, the separation of which at a moment evidently critical indicated an incaution which was doubtless responsible for the exposure also of the main fleet to torpedo attack. Such observed incaution is itself a valuable factor in a military decision. The various facts here given, with the corresponding elements on the Japanese side, stated in a succinct, orderly manner, constitute a problem of exactly the character hypothetically assumed in a war-college problem. When stated, the query follows: Estimate the situation; decide your course of action, which is styled, technically, the Decision; and for its execution formulate your orders to subordinates. The orders to each subordinate will state clearly the situation, the part assigned to himself, with as much information concerning the movements of others engaged in the general operation as will, or may, enable him to act intelligently. What the subordinate is to accomplish—his Mission—is made perfectly clear. How he is to do it is left to his own judgment, partly because the circumstances under which he may have to act can rarely be foreseen, chiefly because reliance can be felt that men brought up with a common vision will do the right thing.

At the war college, the propounding such a problem as the one just cited has been preceded by a course of lectures by men whose previous study and experience have constituted them experts. Each officer under instruction submits two papers: (1) Estimate of the situation, deduced from all the factors, at the close of which is formulated a proposed course of action, which is called the Decision; (2) an order, or set of orders, for putting the decision into execution. The estimate of the situation involves, as a factor, a determination of the proper strategic end to be accomplished; the *ultimate* achievement of which end, whether at once or later, is styled the Mission. Upon this follows consideration of the numbers and disposition of the enemy's forces, and of one's own, as modifying the possibility of immediately accomplishing the mission. Thus Mission defines the end; Decision, the practicable first step. If objection be taken to terms such as mission and decision—or doctrine—the reply is that in all technical treatment technical terms are necessary; and that, when once comprehended, they facilitate discussion, exactly as each foreign word acquired facilitates conversation.

For executing the decision, orders are addressed to each subordinate for his particular part in the combination which the decision requires. Both estimates and orders are then reviewed by the instructor, with criticism and suggestion. Ultimately there is a general discussion among all in full conference. Besides the elucidation which any matter receives from the deliberation in common of several minds, this discussion reacts upon the men engaged. It tends to correct errors, yes; but the great advantage is that principles and illustrations enter into the mind more and more through repetition, not only in the particular discussion of the varied phases of a single case, but by reiteration in many discussions of many cases. For a principle, if correct, cannot but recur repeatedly, steadily deepening its grip. Similarly, reiterated instances of disaster from specific dispositions emphasize warning and give security against errors of like general character.

The value of such a study as that suggested above for the Japanese is still more recognizable, if we imagine it undertaken by the Russian staff a year before the war began. This will illustrate the vital connection between national policy and military preparation. Upon this the war college strongly insists, and most properly has embodied in its course. International policies is one of the subjects of study. In the United States people are singularly oblivious of the close relation between peace and preparation. Outside of a few officials of the Navy Department, public opinion about naval development does not take into its reckoning any digested consideration of our international exposures. Granting that the Russian officials kept such account as they should of Japanese military and naval preparation, they would have in hand a year before the war the following data: The size, constitution, and disposition of the Japanese army; the numbers and character of the Japanese fleet; the means of transportation available to Japan. As matters of serious dispute existed, these data constituted elements in the problem: how to follow the national policy and yet maintain peace? The Japanese maritime transportation was a large part of the logistics of Japan, as the Siberian Railroad was of that of Russia. The data mentioned, together with the numbers and disposition of the Russian fleets and armies, formed the elements of a problem; to be formulated by a clear and succinct statement of each and all of the factors named. The same demand follows: estimate the situation; formulate your measures to assure peace or to encounter war (which in such a case are identical); issue the orders necessary to execute the measures. If the estimate of the situation had been undertaken by officers with a competent doctrine, the Decision must have been to strengthen the fleet in the Far East; not by vessels proceeding singly—as was done—but by a division as strong as the Baltic ports could send. An estimate of the situation could not but have shown that, although the Russian navy in the aggregate was superior, the division in the Far East was not as strong as, for security, it should be. The whole navy had been divided injudiciously; the first requisite was to reunite it by measures strategically sound, which the despatch of a string of single ships proceeding out was not. The strong naval conviction prevailing in the United States against dividing our battle-fleet between the Atlantic and the Pacific was derived from the war games of the college, testing the strategic situations resulting from such division.

The war game, which has been used for many years at the war college, attacks the same class of problems as does the written "estimate of the situation" and formulation of measures just described. In it the men who write the "estimates," etc., are pitted one against the other as opponents. Similar data are furnished to each: a statement of the conditions as far as known to his superiors—namely, the disposition of the forces on his side, and such account of the enemy's as may be reasonably assumed to be ascertained, but necessarily less full than that of one's own. Each receives also, as from a national government, general instructions, indicating the particular service expected of his command. This is his "mission": *what* he is to do, not *how* to do it. The place of a chess or backgammon board is taken by a large map embracing the scene of operations, upon which are arranged and moved tokens, representing the positions held by both sides, as well as the numbers and successive dispositions of the various forces. The game thence proceeds, move by move. The two contestants occupy separate rooms, while in a third is an umpire who pronounces on each move; whether, by the experiences of war, it is feasible and so permissible. Within a certain range he decides by his own judgment and accumulated experience, while in other cases there are fixed rules and fixed values assigned to different forces and to different situations. Doubtful cases are under certain conditions submitted to the decision of the dice; thus recognizing Nelson's saying that some allowance must be made for chance, and Napoleon's that war cannot be made without running risks. The game as described embraces all the operations of a campaign, from the start from the base to the collision of the fleets. It thus opens with strategy, which embraces the whole field; narrows gradually till the fleets feel each other's proximity, and are, as it were, manœuvering for advantage on the field, a phase called stratego-tactical; finally, there are the sighting each other and the manœuvers of battle, technically styled tactical. In these last, on the game-board, the rules governing "values" are grounded entirely on the scheme of battle exercises of the battle-fleet in April, 1911; a circumstance illustrating the interconnection between fleet and college, which it may be hoped will be continually greater.

If a nation possesses military positions abroad, many cases in war, and many hypothetical cases at a war college, will present situations which involve both land and sea forces. This added condition constitutes a more intricate problem; but the method of dealing with it, whether by written estimates and subsequent discussion, or by war game, is the same. Owing to more numerous data, the condition is more complex; but the manner of solution will be like.

It will also readily occur that in every war college—and many nations now possess them—the scenes chosen for hypothetical cases to be discussed and solved will be primarily the regions in which general national policies, or particular international relations, make military or naval operations most probable. Historical incidents, *wherever* occurring, are profitable for instruction, for the elucidation or confirmation of the great universal principles of military action; but, for application of those principles, the scenes first to be selected are those where the national forces are most likely to act.

The treatment, purposely discursive because intended to be popular, has made mention of logistics, strategy, tactics, and national policies, giving at greater or less length the character of the subjects thus named, their relation to the purpose of the Naval War College, and the method of treatment; emphasizing the great object of evolving a common mode of thought, and a common appreciation of proper military conduct, among all the officers of a navy. There remains one other subject, International Law. In a country full of lawyers and politicians, with a government possessing a President, Secretary of State, and a large corps of ambassadors and foreign ministers, it may be asked doubtfully why naval officers should give time to international law. The reply is that in this extensive system of functionaries the naval admiral or captain is incidentally one; and that, in international law, as in strategy and tactics, he must know the doctrine of his country. In emergencies, not infrequent, he has to act for his superior, without orders, in the spirit and manner his superior would desire. If in war, the war may be complicated by a dangerous foreign dispute arising from action involving neutral rights; or, on the other hand, a neutral unright may be tolerated to the disadvantage of the national cause. In peace, injudicious action may precipitate hostilities; or injudicious inaction may permit infringement of American rights, of persons or of property. The treatment of international law, consequently, is the same as of the more distinctively military subjects—a competent lecturer and lecture system, the posing of problems, their solution by the student, comment and criticism by the teacher, discussion in full conference.

CHAPTER XI

BRITAIN & THE GERMAN NAVY

THE HUGE DEVELOPMENT OF the German Navy within the past decade, and the assurance that the present rate of expenditure—over £20,000,000 annually—will be maintained for several years to come, is a matter of general international importance. Elsewhere, and in another connection, I have had occasion to point out, in the American Press, that the question immediately raised is not what Germany means to do with this force, which already is second only to that of Great Britain, and for which is contemplated a further large expansion. The real subject for the reflection of every person, statesman or private, patriotically interested in his country's future is the simple existence present, and still more prospective, of a new international factor, to be reckoned with in all calculations where oppositions of national interests may arise.

From this point of view it is not particularly interesting to inquire whether Germany has any far-reaching purposes of invading Great Britain or of dismembering her Empire; nor yet whether, on the other side of the ocean, she purposes no longer in future contingencies to show that respect for the Monroe doctrine which she hitherto has observed, much to American satisfaction. Americans, while giving full credit to

The Daily Mail, 4 July 1910.

Germany for the most friendly intentions towards them, have to note that in the future she can do as she pleases about the Monroe doctrine, so far as our intended organisation of naval force goes, because she will be decidedly stronger at sea than we in the United States expect to be, and we have over her no military check such as the interests of Canada impose upon Great Britain.

THE RIGHT ATTITUDE FOR GREAT BRITAIN

Similarly, the people of Great Britain should not depend upon apprehension of Germany's intentions to attack in order to appraise their naval necessities and awaken their determinations. Resolutions based upon such artificial stimulus are much like the excitement of drink, liable to excess in demonstration, as well as to misdirection and ultimate collapse in energy, as momentary panic is succeeded by reaction. Unemotional, businesslike recognition of facts, in their due proportions, befits national policies, to be followed by well-weighed measures corresponding to the exigency of the discernible future. This is the manly way, neither over-confident nor over-fearful, above all, not agitated. Of such steadfast attitude, timeliness of precaution is an essential element. Postponement of precaution is the sure road to panic in emergency. An English naval worthy of two centuries ago aptly said, "It is better to be afraid now than next summer when the French fleet will be in the Channel."

In this characteristic of precautionary action a democracy like that of Great Britain stands at a grave disadvantage towards a people like the German, accustomed to a strong Government. A German writer[1] has said recently, "In Germany we hold a strong *independent* Government, *assisted* by a

1. Hans Delbrück, *Contemporary Review,* October 1909, 406. My italics.

democratic Parliament, to be a better scheme than the continual change of party rule customary in England." This was substantially the view of James I. and Charles I. in England, and we know what came of it; but it is the German position to-day. Few Englishmen or Americans will accept it; I certainly do not; but for the organisation of force in the hands of a capable Government, such as that of Germany has shown itself hitherto to be, the scheme is much more efficient, because the plain people of a parliamentary country—the voters—refuse to think about international or military matters. Yet it is they who make and unmake Governments, now one party, now the other; and the Government's outlook upon international preparation is always qualified by a look over the shoulder at the voters. This is much less the case where the people have behind them the tradition of being disregarded comparatively. True, no Government, not the most autocratic, can wholly disregard national feeling. The question is one of more or less; and as between Germany and Great Britain, Government in Germany is, as Government, much more efficient for organised action, even though it make less for the kind of development which follows personal freedom and constraint.

THE NAVY BECOMING LESS POPULAR

This is the fundamental condition which the British democracy of to-day has to recognise as regards their national security, upon which their economic future—their food, clothing, and housing—depends: that they stand face to face with a nation one-fourth more numerous than themselves, and one more highly organised for the sustainment by force of a national policy. It is so because it has a Government more efficient in the ordering of national life, in that it can be, and is, more consecutive in purpose than one balanced unsteadily upon the shoulders of a shifting popular majority. Fortunately for Great Britain the popular tradition of the national need for a great Navy still supplies to some extent and for the moment a steadying hand; but to one following from a distance the course of British action in late years it certainly has seemed that this conviction is less operative; that its claims to allegiance are less felt and more disputed. Yet, in case of national reverse following upon national failure to prepare, it is the democracy, the voters, who will be responsible; the voters also who will suffer.

The prolonged formal peace which Europe has enjoyed for thirty years affords a precise illustration of the ineffectiveness of populaces to realise external dangers. Continuance of peace induces a practical disbelief in the possibility of war, and practical disbeliefs soon result in practical action, or non-action. Yet observant men know that there have been at least three wars in this so-called period of peace; wars none the less because no blows were exchanged, for force determined the issues. The common phrase for such transactions is "the risk of war has been averted." The expression is dangerously misleading, because it is supposed that the controlling element in this conclusion has been the adroitness of statesmen, whereas the existence and calculation of force have been really determinative. Force, too, not merely in the raw material, but the organised force of armies and navies ready—or unready—to move. "I had thought," wrote the American General Sherman, "that the War of Secession was settled by the armed forces of the nation, but at a recent dinner of lawyers I have learned that it was done by the courts."

THE WEAKNESS OF INSULAR COMMUNITIES

Such misconception is peculiarly liable to arise in communities insular by position like Great Britain, or remote from the great nations of the world as is the United States. The measure of security from external aggression which such conditions confer—"the water-walled bulwark" of Shakespeare—favours greatly that free internal development for which democracy is probably the most effective of instruments. But the sense of this security, removing the pressure felt by less happily situated peoples, begets an optimistic attitude towards external dangers, fostering unreadiness for war at the same time that it lessens dependence upon organised government. Other national qualities being equal, Continental frontiers promote the establishment of government effective for external action. As we all know, the Roman democracy illustrated this fact by the institution of the dictatorship for emergencies.

For these reasons insular democracies are lax and inefficient in preparation for war, and in natural consequence their wars have been long and expensive. But wars in the future cannot be long, though they may be expensive; expensive of much besides their immediate cost; expensive in advantages lost and in indemnities exacted. Democracies can no longer afford to neglect preparation, relying upon their strength of endurance and faculty for recovery which probably may exceed that of less free institutions. The time for recovery will not be conceded to them any more than it is by a capable general to a routed foe. The only provision of time for recovery open to modern conditions is the time of preparation.

What reason is there in the nature of things that the British democracy should not maintain an Army proportionally as great as that of Germany? None, except that the British democracy will not. The national wealth is vastly greater; but notwithstanding this, which indicates not only a certain greater power but a much greater stake, the national will so to prepare does not exist. Many distinguished Englishmen advocate measures tending to this result—to the nation in arms; but I doubt if anyone outside of Great Britain expects to see it.

There remains the Fleet; and it is the privilege of insular democracies that they can pursue the quiet tenor of their way behind the bulwark of a fleet efficient in numbers—that is, in great preponderance—as well as in intrinsic worth. But note that a State thus favoured is militarily in the same position essentially as one that hires an army of mercenaries. The only difference is that the seamen are fellow citizens: an immense distinction, it will be granted, but it does not invalidate the fact that the mass of citizens are paying a body of men to do their fighting for them. It follows that the least the mass can do in self-respect as for security is to pay amply and timely for the efficiency of the body they thus employ. If they do not pay "with their persons," as the French say, they should with their cash. But the only adequate payment is timely payment—preparation.

Democracies have had various tasks thrown upon them at various times, but never perhaps one equal in difficulty to that which confronts the democracy of Great Britain. As it now stands, the British Empire territorially is an inheritance from times not democratic, and the world is interested to see whether the heir will prove equal to his fortune. There are favourable signs; one of the most so that has met my eye has been the decision of the Labour Government in Australia that in time of war the Australian Navy should be at the absolute disposal of the British Admiralty. Such sentiment, realised in commensurate action, is effective imperial democracy. But my reading has not found the corresponding reflection of this determination in the British Labour Party at home; rather, it has seemed to me, a disposition to undervalue the necessity of preponderant naval force even in European waters.

The security of the British Empire, taken as a whole with many parts, demands first the security of the British Islands as the corner stone of the fabric; and, second, the security of each of the outlying parts. This means substantially British control, in power if not in presence, of the communications between the central kingdom and the Dominions. This relation is essentially the same as that of a military base of operations to the front of the operations themselves.

THE NEW GROUPING

In the present condition of Europe the creation of the German Fleet, with its existing and proposed development, has necessitated the concentration in British waters of more than four-fifths of the disposable British battle force. These facts constitute Germany the immediate antagonist of Great Britain. I do not say for a moment that this manifests Germany's purpose; I simply state the military and international fact without inference as to motives. The geographical situation of the two States reproduces precisely that of England and Holland in the early days of Cromwell. It was not till the nations had fought and the Dutch were reduced, less by battle than by trade destruction, that the relief of pressure in the North Sea enabled English action abroad. This result was attained more satisfactorily forty years later by the alliance of the two States under the impulse of a great common danger; but whether that alliance would have been feasible without the antecedent settlement by trial of strength is disputable. In the course of the earlier war the Mediterranean was abandoned by the English Navy in order to concentrate in home waters, and this concentration, coupled with the commanding position of the British Islands with reference to Dutch trade routes, determined then the issue.

The British Navy to-day has in great degree abandoned the Mediterranean for a similar concentration. Over four-fifths of the battleship force is in the "Home" and "Atlantic" divisions. The Mediterranean has fallen from eleven battleships in 1899 to six in 1910, and these six are of distinctly inferior power. What is the contemporary significance of this fact reproductive of a situation near three centuries ago? Constitutive, too, of a situation now novel; for during more than two centuries British preponderance in the Mediterranean has been a notable international factor. The significance, as read by an outsider, is that in the opinion of the Government, under present conditions of preparation, the security of the British Islands requires the weakening, almost to abandonment, of the most delicate, yet very essential, link in the system of communications of the Empire.

It is entirely true that for the moment the naval concentration at home, coupled with the tremendous positional advantage of Great Britain over German trade routes, constitutes a great measure of security; and, further, that the British waters, occupied as they now are, do effectually interpose between Germany and the British oversea Dominions. The menacing feature in the future is the apparent indisposition and slackness of the new voters of the last half-century over against the resolute spirit and tremendous faculty for organising strength evident in Germany.

THE FUTURE PERIL

An examination of present and probable future European international relations is plainly incompatible with my space; but speaking as an onlooker, studying these, and following the tone as well as the words of parliamentary debates, I have thought to see the growth of a spirit which threatens to leave Great Britain unprepared to hold her own and to sustain her Empire in the very probable contingencies ahead. Impelled to weigh these seriously, the impression has gained ground, against a steady previous conviction, that Great Britain would prove equal to her fortunes.

In a recent American magazine[2] a German writer, reported

2. Theodor Schiemann, "The United States and the War Cloud in Europe," *McClure's Magazine*, June 1910, 223.

to be a trusted confidential friend of the Emperor, has said, "The weak man cannot trust his judge, and the dream of the peace advocate is nothing but a dream." The concentration of the battle fleet in home waters is correct; the relative abandonment of the Mediterranean for that purpose, if for the moment only, is likewise correct, especially as the "Atlantic" fleet may be considered an intermediate body, a reserve, able to move eastward or southward as conditions require; but the clear reluctance to acquiesce in present naval requirements is ominous of a day when the Mediterranean may pass out of the sphere of British influence, centred round the British Islands exclusively. This will symbolise, if it does not at once accompany, the passing of the Empire, for a hostile force in the Mediterranean controls not only an interior line—as compared with the Cape route—but an interior position, from which it is operative against the Atlantic as well as in the East.

It is difficult to overstate the effect of this upon the solidity of the Empire, for the Mediterranean is one of the great central positions of the maritime world. A weakened Mediterranean force is the symptom that neither as principal nor as ally may Great Britain be able to play the part hitherto assumed by her in the great drama of which the awakening of the East is the present act; while among the *dramatis personae* are Egypt, India, Australia, and New Zealand.

CHAPTER XII

THE PANAMA CANAL AND THE DISTRIBUTION OF THE FLEET

THE QUESTION of the proper distribution of a national navy is not only of great importance, but often of much perplexity to a State having large external interests; especially if these be not only extensive, but divergent. Great Britain, despite her enormous fleet, has for two centuries past illustrated this, owing to her wide-spread commercial system and scattered dependencies. To use the strong expression of a French admiral, "In the midst of riches she has felt all the embarrassment of poverty." The British official who answers to our Secretary of the Navy wrote to the celebrated Rodney, "It is impossible to have a superior fleet in all quarters." Upon this followed the corollary, that the fleet must be ready to move in force from one quarter to another, according to the turn of the struggle; a readiness which can be perfectly assured only by keeping it together. The difficulty, in short, is one that cannot be removed entirely, because the causes cannot cease to exist; but it can be met with good prospect of success, provided well-settled principles, based upon past experiences, are duly and steadily observed in practice.

Fundamental among these principles is that of concentration, a word which may be said to include the whole of military art as far as a single word can, as it comprises also the

North American Review, vol. 200 (September 1914), 406–17.

secret of successful purpose in any enterprise and in any calling. But concentration is a general term, the application of which is determined by the specific circumstances of each case. Of such circumstances, position is among the most decisive. War, said the great Napoleon, is a business of positions. The point of concentration, as well as the necessity for it, has to be considered. Concentration itself might be considered a species of position, in that it decides that the position of the fleet shall be single, not dual.

It is purposed here to apply these remarks to the case of the United States Navy, under the conditions consequent upon the completion of the Panama Canal. One first essential to be noted is that any general disposition adopted should have direct reference to a state of war, and as far as practicable should conform to that which the opening of war requires. Independent of the fact that such an arrangement accelerates mobilization, there is the further very important consideration that a change of dispositions, when political relations are strained, may, by the impression produced on another government or people, precipitate the very issue which diplomacy is seeking to avert. Since these words were written, the persistence of Russia in mobilizing—not any hostile action on her part—is alleged by Germany as her reason for declaring war. Not military readiness only, but sound civil policy also, dictates that the dispositions of peace anticipate the demands of war.

The case of the United States, with two seaboards so widely separated in water distance as the Atlantic and Pacific, is not unprecedented. It is only an extreme instance of conditions found elsewhere. Spain, and still more France, have known inconvenience, and at times have experienced disaster, from the division of naval force between the Mediterranean and the Atlantic. Constrained by interests on both coasts, and by the administrative necessity of providing navy-yards on both, because either might be the chief scene of a war, the fleet was distributed between the two. The effort subsequently to concentrate, whether in one home port or at some external position, led to many strategic mishaps, entailing at times not only failure, but destruction. Trafalgar is a signal instance of a massive catastrophe, the prelude to which was a series of abortive attempts to combine several squadrons previously divided between the two seaboards of France and Spain.

An example of the same, more striking to us because contemporary, was the fatal policy which led Russia in her recent war with Japan to dally with concentration, and to permit her fleet to remain divided between the Baltic and the Far East, while peace still existed. Two years before the war began, the larger part of the Baltic fleet was already in the Far East, in force substantially equal to the Japanese Navy. But it might have been superior; and the practice of sending reinforcements in detachments enabled Japan, noting the course of events, to declare war at the critical moment when one was on the way which might turn the scales. It could not proceed, because the Japanese fleet barred the junction, and Russian equality was prevented from becoming superiority. The effort to concentrate at an improper position, instead of assembling in home waters and proceeding thence together, drew the jealous attention of the enemy, who not only was enabled, but necessitated, to strike before the meeting was effected. A consummate master of the art of war, commenting on a similar military conjuncture a century before, wrote, "What complicated pains to concentrate in the face of the enemy, when it could perfectly well have been done beyond his reach!"

Undoubtedly, conditions arise which necessitate division of effort. For example, Great Britain is compelled now to a concentration of war shipping in home waters, more imposing than any she has had to maintain since the navy of Holland rivaled that of England two hundred and fifty years ago. The cause, too, is the same; substituting the German Empire of today for the Holland of the seventeenth century. But while this position of the British main fleet covers, as against Germany, all approaches to British shores by the Atlantic, it does not equally guard routes using the Mediterranean, whether for commerce, or for access to political interests in Egypt and India. These remain exposed and must be protected; for a very large fraction of British trade originates in the Levant and Black Sea, while still more comes from the Far East, passing by the Suez Canal through the Strait of Gibraltar. Either by her own power or by secure alliance, it is essential to Great Britain so to control the Mediterranean that her communications throughout should be safe against the possible action of Germany's Mediterranean allies, Austria and Italy.

When such widely divergent yet indispensable interests are at stake, there are two principal means of defense. One is to be superior on both scenes; the other is such a distribution of aggregate force as to give a probable chance of concentration in superior numbers at the point where danger is imminent, before the enemy can himself act. It is evident, however, that for Great Britain no distribution is permissible which will deprive her home waters of superiority over a possible antagonist so near as Germany. She is compelled to the first alternative,—superiority on both scenes, either by her own ships or by those of an ally.

It will be noted that the nearness of Germany to Great Britain herself, and that of her allies to British vital interests in the Mediterranean, form a combination of simultaneous perils, constituting a peculiarly menacing situation. If Great Britain were equidistant from both, she would have a central position, which might afford opportunity to meet first one and then the other, in successive encounter; while if there was little probability that war would spring up in both quarters at the same time, conditions would be still further modified. But in each case concentration, to the extent of assured superiority at the point of contact, is the one thing needful. That such concentration should be the controlling factor in peace dispositions,—should be the normal state then,—is evident from the rapidity with which modern wars develop, and from the political fact that, when relations are strained, significant movements may precipitate hostilities.

Whatever change in international relations the remoter future may have in store, it is fairly sure that the present outlook makes improbable any conjuncture of simultaneous dangers for the United States, in both Atlantic and Pacific, such as hangs over Great Britain in her home waters and the Mediterranean. If Panama be held securely, no one naval enemy can threaten both our coasts at the same time, without great and undue risk to itself. Concerted action to the same end, by an Atlantic naval power in co-operation with one in the Pacific, is unlikely. This may be inferred from the terms of the treaty of alliance between Great Britain and Japan; and still more from the apparent acquiescence in the general principles of the Monroe Doctrine on the part of the naval states of Europe. Although not formulated, this acquiescence has been shown very practically in more than one connection; notably in the still pending Mexican troubles. Europe indeed has in the Balkans, in Asia, and in Africa, preoccupations so critical as to disincline any single state from embarking in a policy of American adventure.

Nothing of this, however, modifies the policy of concentration, in the sense that wherever the fleet may be at a given moment it should there be in local superiority to any probable enemy. The Russian navy, being superior in the aggregate to that of Japan, it was of comparatively little importance whether it was concentrated in the Baltic or in the Far East; but it was of immense importance that it should be concentrated, not divided. As a general proposition, this evidently implies more than the formulation of a mere strategic requirement. It applies equally to national naval policy: that the navy, as constituted by legislation, should be big enough to assure such superiority. Naval policy is essentially and supremely a question of foreign relations.

Granting a superiority based upon properly calculated estimates of international relations, concentration of the battle-fleet of the United States is a matter of much more consequence than its precise position. Not that position is of less than great importance. The Russian battle-fleet would have been much better placed in the Far East than in the Baltic. It was not adequate to both, as the event proved; but, if it had been united, its remoteness—in the Baltic—would not have occasioned the decisive disaster which division entailed. It was the business of the Russian Executive to form its estimate of the general European situation, including therein its own secret purposes; and then, before war threatened, to assemble its battle-fleet, and send it where it should be most surely at hand, if war came; but on no account to divide it.

In point of distance, the Baltic and the Far East constituted a dilemma not very unlike that of our Atlantic and Pacific seaboards before the Canal is open for use. The Canal completed, and secured against hostile enterprise, we shall have there a central position, similar to that imagined above for Great Britain as to Germany and the Mediterranean. Even before the Canal, however, despite the immense distance and the administrative difficulties, of coal and supplies, connected with transferring the fleet from ocean to ocean, the dictates of sound policy demanded the concentration of the fleet, not its division between the two; for the plain reason that the margin of superiority was then, and is now, not large enough to permit separation. Halve the fleet, and it is inferior in both oceans. Divide into unequal fractions, keeping in one a bare superiority, and you have in the other a detachment in itself adequate to nothing except to soothe the tremors of old women and of the childish on shore; tremors of the character which lowered rents on the south shore of Long Island during the Spanish War, because of apprehension that an enemy's ships might spend (waste) ammunition on an open beach; whereas, joined to the main body, such a reinforcement may constitute a superiority so decisive as to prevent war. It is to be remembered also that the nominal aggregate of a military force is rarely available under the stress of actual war. The "present for duty" of troops in the field represents usually considerable reductions by sickness, detachments, and other incidents of service. So in a fleet, reliefs, accidents, detachment for repairs or for recreation of the crew, cause deductions for which a margin of allowance must be made in calculation.

The Canal modifies the previous situation by minimizing all the difficulties of transfer, but it does not change the dictate as to concentration. Even if Panama were a natural waterway, like the Strait of Gibraltar, an enemy by occupying it in force would acquire advantage for keeping apart the divisions in the two oceans, if not already united. But an artificial channel, with locks, in a region like Panama, stands always in risk of interruption. Accident, surprise, treachery, a momentary lack of vigilance, other fortune of war, may effect a prolonged block of an essential line of communication, affording an enemy a strategic opportunity, through possessing decisive local superiority for whatever the period of closure may be. The provision against this is concentration. It may happen to be on the wrong side of the Canal at a critical moment; but it is better that such moment should find all on the wrong side than only half on the right, because transfer is always more feasible than junction, and the half might be annihilated while the whole could not.

The people of the Pacific coast have shown themselves from time to time sensitive, if not apprehensive, about the absence of the main fleet from their shores. They have felt themselves to be the more endangered, both by position and by the smallness of the resident population as compared with the Atlantic seaboard. This is true; and upon it they have based a claim for a proportion of the battle-fleet to be stationed in their waters, thus dividing the force, and that under conditions of very great exposure. On the other hand, the Atlantic coast communities feel the claim of greater numbers,—the claim of a majority; reinforced, of course, by the inevitable superior national concern in the larger commercial, manufacturing, and other interests, which superior numbers accumulate. On their side also is long-standing tradition. Men have not yet adjusted their thought to the new condition, that the Pacific rather than the Atlantic holds the problem of the near future; that Europe and Atlantic America have reached fairly stable conditions, both in themselves and toward each other; and that both are looking outward, the one eastward, the other westward, toward Asia. Further, questions of administration, of supply and repair, are facilitated by the greater development of navy-yard equipment, occasioned by the hitherto usual presence of the ships in the Atlantic. These local feelings are an inevitable attendant of human nature. They carry with them the evil of sectionalism, and constitute a problem for the government, which in a democracy has to have regard to votes. The one solution, and the perfectly adequate reply, is that a military question, in this resembling all technical questions, must be settled on technical grounds; in this case on military grounds.

In naval matters, however, international relations form a part of the military problem; and while these cannot modify the requirement of concentration, they do affect the questions of position and of the necessary numbers of the fleet. Consequently, while every naval officer who respects himself and his profession should be well informed as to international conditions, for not otherwise can he form sound military judgment or give adequate counsel when called upon, the general decision as to position belongs primarily to the civil government in its executive branch; for, besides its control over the military services, it is charged with the ultimate responsibility of action, and it alone necessarily possesses the needed information. A very critical part of this knowledge is the actual state of negotiations at any moment, the temper of other governments, and of their people; upon which depends the policy of fleet movements which might be construed to indicate distrust or offensive intention. All this responsibility is civil and executive; the military adviser may contribute sound military opinion, but decision rests elsewhere; in last analysis upon instructed public opinion.

From this consideration springs the desirability of maintaining generally such dispositions as correspond to the demands of opening war, and which from their general permanence have no particular significance at a given moment. An instance of the contrary may be recalled in the answer of the British government to the German Emperor's telegram to Kruger. An additional squadron was ordered into commission—a perfectly pronounced diplomatic utterance. However timely in the particular emergency, such action may be very untimely at another, and yet indispensable to safety. This dilemma should be forestalled.

A permanent arrangement of the character denoted would be that of planned frequent interchange of the main fleet from coast to coast. As far back as 1907, when the battle-fleet made its voyage to the Pacific by way of Magellan, and ultimately round the world, I suggested the periodic repetition of the transfer; as tending not only to general efficiency, but to increased aptitude in the administrative processes involved. To this, in my judgment, should be added a practical recognition that the Caribbean Sea and the Panama Canal form together a great central position, corresponding to one before imagined for Great Britain, and the most important within the sphere of action of the United States. This can be done either by designation,—the Caribbean Fleet; or by customary presence there, as being the center, to and from which movement takes place. One effect of this, and of the interchange advocated, would be to enforce the necessity for developing dockyard equipment and supplies both in the Caribbean and in the Pacific; now less complete than they should be from the military point of view. Another gain would be the facility which practice gives in passing the fleet from sea to sea. Although the manipulation must be always under the charge of the Canal force, it is likely that, as a military measure, repetition would develop methods in the management of the fleet conducive to rapidity and security. A right of way for the whole fleet, unbroken by merchant vessels, should be guaranteed.

Above all, as a political measure, interchange would tend to appease sectional jealousy; while the assumption of the Canal and Caribbean as the main habitual station of the fleet would recognize actual international conditions, and in military calculations would form a sound habit of mind, which is possibly even more important than correct position accidentally taken,—not based on reasoned judgment.

The subject has been treated so far from the merely defensive side; from the standpoint only of national local security as involved in the distribution of the fleet. It is to be borne in mind that this may have larger—or, rather, wider—functions to perform. It may be thought necessary, even from motives of defense, to transfer the fleet to external possessions, such as Hawaii and the Philippine Islands. Under particular circumstances it may be considered that defense is best promoted by offensive action; for such action, judiciously planned and adequately executed, tends to keep the enemy's fleet where our own is, and therefore distant from our shores. "It is suggested," wrote Nelson in 1801, "that the Danish fleet will take advantage of our going up the Sound to escape and get to France; but I own I do not think they will send away so large a force while their capital and their own shores are threatened." In the days, centuries ago, when England really feared Spain, her seamen thought an attack on Cadiz, or elsewhere near the Spanish heart, the surest means to secure English shores. "Singeing the King of Spain's beard," they called it.

This is but a commonplace of military art, and of the experiences upon which that art is founded. Napoleon in 1812, having in view the protection of Badajoz, then a French fortress in Spain, wrote as follows to Marmont, commanding in that region: "Concentrate your army around Salamanca (over a hundred and fifty miles north of Badajoz), keeping ready for instant action. There you are master of all Wellington's movements. If he undertakes to march upon Badajoz, let him go. March straight upon Almeida (a principal British fortress, seventy miles west of Salamanca), and you may be sure he will quickly return. But he understands his business too well to commit such a fault [as making such an attempt] with you by your position threatening Almeida." During our hostilities with Spain, in 1898, while Cervera's squadron was still in Santiago undestroyed, a Spanish division under Cámara sailed hurriedly from Cadiz for Suez, apparently intended against Dewey in the Philippines. It passed the Suez Canal two days before Cervera left Santiago. The reply was to detach an American squadron of adequate force to operate against the Spanish coast. The squadron did not sail, but the publicity of the measure would tend to prevent Cámara from going farther; the more so that the whole American fleet was liberated for similar action so soon after, by Cervera's defeat. It will be noted also, that Cámara and Cervera, having been separated by the primary dispositions of Spain, could not afterward unite.

Security, therefore, is not always, nor most certainly, attained by the immediate presence of the defensive force at the position to be defended. Often the purpose may be better accomplished by action elsewhere. It follows that this contingency also must be contemplated in the peace distribution of the Navy. Independently of the greater efficiency which usually characterizes a large assembly of vessels, owing to the mere stimulus of numbers and to the wider mutual competition thereby induced, the fleet when concentrated on the one coast or the other will be more quickly ready for action, sooner and more effectually mobilized, than if separated. To concentrate is more difficult than to disperse, and all administrative processes also will be hastened. It is true, doubtless, that on a coast properly provided with yards preparation is expedited by distributing the fleet among them; and if preparation has not been completed, as it should be, before war comes, such division will be necessary. Concentration, however, in the military sense, does not mean always the immediate contact of the units concerned. Napoleon's instructions to Marmont, quoted above, assigned for the several corps two marches from Salamanca, the center of movement, as a concentration adequate for the particular purpose; because, so distributed, actual junction could be effected speedily enough. "Supporting distance" is the technical expression. So a fleet may be safely dispersed among navy-yards, provided conditions are such that preparation can be made and junction effected before a concentrated enemy appears. All this, however, is better done before war can begin; while, if it has been postponed, the whole process, if the fleet is together, can be completed sooner than if at the outset part is upon the Atlantic coast and part on the Pacific. As no useful end can be accomplished by such division, there appears no valid military argument against sustained concentration in peace.

In conclusion, a word may be said as to the real military relation of the Canal to the Navy, and of the Navy to the Canal; a question not always understood, and thought by some to have a relation to the distribution of the fleet. It is said at times, somewhat querulously, that when there was no Canal this was advanced as an argument for a larger fleet, both coasts needing naval protection; but that when the Canal became an assured certainty its protection was alleged in turn as a reason for increasing the Navy. Some eminent citizens, a few years ago, memorialized Congress against fortifying the Canal, because, "with all the fortifications possible, it is still apparent that in time of war a guard of battle-ships at each entrance would be an absolute necessity, and equally apparent that with such guard the fortifications would be unnecessary." It is not easy to cite a more egregious instance of the dangers of the ignorant dealing with technical questions.

The relation of the Canal to the Navy is that it opens a much shorter line of communication between the Atlantic and Pacific coasts, and thereby does enable a given number of ships—a given strength of fleet—to do a much greater amount of work; in the sense that it is able to reach one coast from the other in so much less time as is required to go by it instead of by the Strait of Magellan. Such an advantage may be represented in terms of fewer ships, as well as of less time. It is conceivable, though not probable, that both coasts might be exposed to attack at the same moment. Without the Canal this contingency could be met only by two fleets—that is, one of competent number on each coast. With the Canal not only is transfer quicker and, as to administrative problems, easier, but a fleet smaller in numbers than the aggregate of the two, yet decisively superior to either enemy, has the chance of destroying first the one and then the other, as the Japanese destroyed first the Port Arthur fleet and then Rozjestvensky's. The value of the time element contributed by the Canal is apparent.

Under present conditions, such a combination of enemies is unlikely, although in calculation it must be contemplated. With but one enemy, the Canal saves time, if the concentrated fleet has to go from one coast to another. In last resort, if properly fortified, the Canal affords a retreat in case of reverse, and a means of speedy return when reestablished. The Canal, in short, is a central position, from which action may be taken in either direction, and it is also a decisive link in a most important line of communications. It is possible that, in the European war that has begun since these lines were written, the Kiel Canal may afford pertinent illustration.

That the Canal may so serve it must be fortified, and able to stand by itself, without battle-ship help against attack. The relation of the fleet to the Canal is that of every fleet to a port that has back of it no immediate resources, and must be supplied from home; for instance, the relation of the British fleet to Gibraltar. The fleet keeps open the communications by controlling the sea. I doubt if during the three years' siege of Gibraltar the navy proper fired a gun in defense of the port; it was there very rarely, at long intervals, to bring supplies. The Russian ships shut up in Port Arthur were equally useless for assistance in the defense. To detach from the fleet—to divide it—in order to assist in defense of the Canal, is not only open to the same objection as division between the coasts, but it will have the further disadvantage of being a measure inherently futile to the proposed end.

The Canal, therefore, assures the communications of the fleet, and in this respect is to be considered as a highway, as a means of transit. The fleet assures the communications, the line of supplies, to the Canal and its defenses, which from this point of view are an advanced base of operations. These services are reciprocal, but distinct. That Panama will have the unique privilege of two entrances, one on each ocean, assuring two lines of supplies, widely divergent, emphasizes its independence, and that of the fleet; which, when acting in one ocean, has thus a covered line of supply in the other. In the matter of defense, regarded as a question of mere fighting, the fleet and Canal have no essential connection with each other. The Canal should be so fortified as to be indifferent, at a moment of attack, whether the fleet is in its ports or a thousand miles away.

INDEX

Académie de Marine, 82, 83
Actium, 14
Aden, 314, 333, 335, 339
Adriatic Sea, 183
Aegean Sea, 199
Africa, 287, 300, 332, 372. *See also,* South Africa; Wars, Boer; Cape of Good Hope
Alexander the Great, 2, 23, 276, 277
Alexandria, 223, 345
Algiers, 102, 185
American Revolution. *See* Wars, and *Major Operations of the Navies*
Antwerp, 161
Applicatory system, 351–58
Armaments and Arbitration, quote from, xxi; chapter X, 343–58
Armor, xxxi, 122, 152
Army, and relation to the navy, 94n, 182, 184, 260
Asia, 372
  and immigration to USA, 174, 175
  See Problem of Asia; "Persian Gulf . . ."
Atlantic Ocean, xx, 175, 188, 198, 235, 296, 297, 299, 316, 356, 367, 369, 372, 375
Attack, xxi
Australia, 158, 160, 315, 316, 317, 328, 332, 333, 364, 367
Austria, 183, 260, 261, 264, 371

Balance of power, 298
Baltic Sea, 32, 34, 40, 64, 67, 149, 265, 296, 297, 299, 332, 356
Bart, Jean, 303
Bases, xviii, xx, xxvi, xxix, 21, 101, 123, 146, 150, 176, 177, 182, 183, 184, 212, 231, 235, 245, 251, 255, 260, 261, 282, 288, 290, 291, 305, 365
  advanced, 244
  assembling forces in, 132–33
  continuing support from, 138–40, 146, 147, 250
  cutting loose from, 189
  launching forces from, 133–38, 225
  permanent, 36, 169, 170, 257
Battle, xxxi, 250, 251, 253, 270, 274, 289, 313, 351. *See also* names of battles; *e.g.*, Trafalgar
Battle-fleet, 104, 300, 302, 303, 312, 313, 357, 365, 376. *See also* Fleet
Beachy Head, battle of, 194, 247
Belgium, 37, 38, 44

Bermuda, 114, 119
Biscay, Bay of, 190, 223
Bizerta, 102
Black Sea, 14, 183, 189, 199, 307, 332, 371
Blenheim, battle of, 234
Blockade, 25, 33, 46, 91–92, 107, 108, 123, 212, 218, 225, 228, 250, 251, 289, 302, 312, 313
Blue Water School, 169, 185, 263
Bomilcar, 15
Bonaparte. *See* Napoleon Bonaparte
Bosnia and Herzegovina, 101
Bosporous, 189
Brest, 136, 161, 187, 222, 228, 229, 263, 300, 302, 309, 312
Britain, 159, 171, 285, 319
  article "Britain and the German Navy," 359–67
  and British naval experience as a guide, xviii, 51, 232
  and character of government, 63–72
  and defensive role of fleet, 268
  and Empire, 158, 310, 315, 317, 364, 366, 367
  and geographical position, 31–36, 108, 291–92, 299, 326, 371
  and Japan, 286, 307, 308, 341, 372
  and national character, 60–61
  and navy in 1880–1914, 123–24, 287, 298, 299, 300, 305, 306, 307, 315, 326–30, 335, 365
  and *Pax Britannica*, xvi-xvii, 101
  and trade, xxx. *See also* Shipping, merchant
  and war with Germany, 144, 155, 262, 263, 336–42, 359–67
Brueys, Admiral, 191, 213
Business, and war as, xxxi
Byng, Admiral John, 157, 251

Cadiz, 186, 187, 228, 229, 235, 327, 333, 378
Caesar, Julius, 2, 23, 276, 277
Cámara, Admiral, 378
Cambridge University, xv
Canada, 158, 171, 172, 175, 232, 333, 360
Canal Zone. *See* Panama Canal
Cannae, 15, 17, 18
Cape of Good Hope, 30, 34, 109, 118, 297, 314, 316, 367
Cape Passaro, battle of, 67
Cape St. Vincent, battle of, 11, 50, 159, 238, 350
Capital, as an objective in warfare, xxvii
Caribbean Sea, xx, 35–37, 106, 107, 118, 138, 171, 173, 175, 178, 179, 183, 185, 186, 229, 243, 253, 254, 259, 264, 305, 376, 377
Caroline Islands, 102
Carthage, 16–22
Cervera, Admiral, 378. *See also* Santiago
Ceylon, 314, 316
Chance, xxvii, xxx, 282, 357. *See also* Risk
Channel, English, 13, 24, 32, 33, 34, 36, 103, 109, 113, 115, 129, 143, 178, 180, 183, 190, 192, 194, 222, 223, 229, 292, 296, 297, 298, 299, 300, 301, 304, 310, 311, 315, 316, 360
Character, national, 53–62, 322
Charles, Archduke of Austria, 105, 149, 235, 262, 267
  quoted, 117, 260–61, 276, 279–80, 352
Chatham, 171, 183

Cherbourg, 300, 309, 312
Chesapeake Bay, 147,
China, 287, 307, 308, 314–17, 330–32
  See also Open Door Policy
Cienfuegos, 129, 146, 178, 257, 258
Ciudad Rodrigo, 220, 221
Civil War. *See* Wars, American Civil; *Admiral Farragut, Gulf and Inland Waters*
Civita Vecchia, 213
Clausewitz, Carl von, xi, 258
  ATM on, 99n.1
Clerk, John, of Eldin, 83
Coalition of powers, 298
Coast defence, xviii, 125, 130, 131, 171, 181, 183–84, 225, 261, 272, 296, 301
Collingwood, Admiral Lord, 350
  quoted, 348
Colomb, Vice Admiral Philip H., xi, xvii
Colonies, 30, 59–62, 88–89, 109, 153, 155, 158, 177, 289, 300, 315, 317, 368
Combined operations. *See* Operations, combined
Command of the sea, xx, 194, 196, 198, 298
Commerce, 283–87, 296
  raiding. *See* Shipping, merchant; *Guerre de course*
Communications, xix, xxvi, xxix, 11, 17, 97–98, 108, 144, 164, 167, 176, 177, 178, 184, 187, 189, 197, 208, 212, 223, 227, 231, 232, 235, 244, 255, 256, 273, 284, 288, 290, 291, 293, 304, 345, 364, 380, 381
  electronic. *See* Warfare, changed circumstances of modern
Concentration, xix, 265, 369. *See also* Fleet, concentration of
"Considerations Governing the Disposition of Navies," 281–318
Contact, division between strategy and tactics, 9
Control, xix, xx, xxix, 1, 15, 18, 178, 179, 180, 181, 192, 227, 232, 235, 239, 260, 292, 294–95, 304, 327, 364, 381
  disputing, 182, 232
Convoying, xxvii, xxix, 185, 186, 187, 188, 192, 193, 194, 196, 198
  See also Shipping, merchant
Copenhagen, 200, 265, 266, 377
Corbett, Sir Julian, xi, xvii, xxxii, 224n.2, 251, 258
Corfu, 209, 211, 212, 214, 215, 223
Corsica, 35, 192, 236, 237, 244, 300
Creasy, Sir Edward, 199
Crete, 115
Cromwell, Oliver, 64–65, 365
Cruisers, xxix, 108, 123, 184, 188, 228, 231, 250, 269, 299, 300, 303, 312, 313
  armored, 301
Cuba, 106, 107, 109, 112, 140, 151, 158, 172, 178, 194, 195, 229, 230, 236, 243, 244, 254
Cyprus, 101

Daily Mail, ATM article from, 359–67
Danube River, 162, 231, 233, 260
Dardanelles, 189, 307, 333
Daveluy, Commander, 273
  quoted, 213, 274–75
Decision, 353, 354
Defensive, xviii, xx, xxi, xxii, 93n.9, 111, 113, 125, 126, 128, 156, 159, 164, 167, 181, 182, 211, 212, 219, 224, 225, 256, 257, 258, 259, 263, 272,

Defensive (*continued*)
273, 274, 288, 289, 294, 295, 304
and dispersion, 265
strength in, 120–31

Delbrück, Professor Hans, quoted, 360–61
Deterrence, xxi
Diplomacy. *See* Statesmen and Statesmanship
Distance, as a factor, 296
Dockyards, 139, 141, 161, 173, 183, 317, 318, 369, 375, 377
Doctrine, 348–58. *See also* Principles, military and naval
Dover, Straits of, 190, 223, 297
Dutch. *See* Netherlands

Education, naval, xiii, 346–47
Egypt, 101, 115, 161, 162, 163, 164, 165, 179, 190, 192, 194, 197, 207–18, 231, 253, 314, 334, 367
See also Nile, battle of
Elliott Islands, 134, 136, 169
Endurance, 120
England. *See* Britain
English Channel. *See* Channel
Estaing, Admiral Comte d', 176
Estimate of the situation, 353–56
Eugène, Prince, 233, 234, 276, 277
Expansion, xxviii, 80, 170
Expeditions, 167, 168, 177–218, 219

Farragut, Admiral David G., xxvii, 209
See also Admiral Farragut
Ferrol, 187, 228, 310
Fleet, xxi, xxii, xxvi, 155, 188, 255, 345, 363–64
composition of, 184
concentration of, 103–5, 146, 197, 225, 311, 365, 367, 368–73, 378–79
disposition of, 281–318, 368–81
divided, 152, 177, 186, 189, 197, 198, 214, 250, 253, 356, 370, 373, 381
neutralizing enemy's, 304
and submarines, xxx. *See also* Battle-fleet; Strength, naval
Fleet-in-being, xxii, 154, 185, 205, 210, 247, 263, 327
Force, xxi, xxii, xxiv, xxv, xxvi, 298
containing, 20n.6
displacement of, 213, 273, 274
distribution of, 103, 189, 281–308
and reserve, 48–52, 139, 367
See also Armaments and Arbitration
Forts and Fortification, xxvi, 119, 122–23, 126, 149, 169, 176, 188, 220, 225, 235, 245, 261, 272, 292, 380
See also Strength, military
France, 101, 262, 305, 307–9, 360
and character of government, 74–87
and geographical position, 31–38, 292, 300
and national character, 56–61, 261
and the navy, 7, 48, 75, 86–87, 287, 300
Frederick the Great, 23, 276
From Sail to Steam: Recollections of a Naval Life, quote from, xxxi

Gage, lee and weather, 6–7
Galley, 77, 118, 203
similarities to the steamship, 2–7
Genius for war, xxvii
Genoa, 236, 237
Geography, and position, 31–37, 109

and extent of territory, 45–47
and physical conformation, 37–45
Germany, 51, 117, 150, 172, 193, 234, 262, 310, 369
and colonial possessions, 102, 109, 170, 337
and geographical position, 27, 34, 114, 163, 260, 296, 310
and naval development, 227, 337–42
and war with Britain, 144, 155, 161, 183, 262, 296, 341, 359–67, 370–73, 376
Gibraltar, 13, 30, 32, 88, 89, 102, 111, 113, 114, 119, 140, 155, 156, 164, 171, 176, 178, 179, 180, 181, 183, 186, 189, 204, 215, 229, 230, 234, 237, 238, 251, 292, 305, 306, 309, 327, 334, 345, 371, 374, 381
Goldsborough, Admiral, quoted, 99
Government, xxiii,
character of, 31, 62–94
influence on sea career of its people, 88–89
Grant, General Ulysses S., xxvii, 156, 209
Grasse, Admiral de, 246
Gravelotte, battle of, 348
Great Britain. *See* Britain
Great Lakes, 235
Great White Fleet, around the world cruise of, 104, 376
Guantanamo Bay, xx, 106, 107, 158, 170, 183, 244, 245, 259
Guerre de course, 33, 108, 187, 365. *See also* Shipping, merchant
Guichen, Admiral de, 7
Gulf of Mexico, 37, 42, 106, 107, 138, 143, 151, 183, 188
Gulf and Inland Waters, xiii
Gunnery and guns, xxxi, 122, 134, 146, 203, 314, 381
Gustavus Adolphus, 276

Haiti, 178
Hannibal, 2, 15, 16, 17, 20, 21, 276, 277
Harbors, 114, 128
Hasdrubal, 20, 21
Hawaii, xx, 102, 117, 119, 158, 170, 172, 174, 175, 190, 192, 377
Hawke, Admiral Lord, 135, 138
Hermocrates of Syracuse, 3n.1, 8–9, 202, 204, 205, 208, 211, 217
History, study of, ix–x, xxiii, xxvii, xxix, 1–3, 23, 276–80, 351, 357
and historical narrative, xii
and lessons of, 4, 9–11, 95–96, 298, 303, 313
Hong Kong, 158, 314
Hoste, Paul, 83
Hotham, Admiral, 247
Howe, Admiral Lord, 9, 347

"Importance of Command of the Sea," quote from, xx,
India, 163, 179, 239, 242, 314, 315, 317, 327, 328, 334, 335, 367
Indian Ocean, 162, 251, 297, 335
Industrialization, xvi–xvii, xxiii
Influence of Sea Power Upon History, 1660–1783, x, xi, xxxii
quote from, xviii, xix, xxiii, xxv, xxvii, xxviii, xxix, xxx
Introductory, 1–26, chapter I, 27–96
Influence of Sea Power Upon the French Revolution and Empire, quote from, xxviii
Inspiration, in war, xxix, xxx

Interest of America in International Conditions, quote from, xxi
Interest of America in Sea Power, Past and Present, quote from, xix, xx, xxii, xxvi, xxvii
Interests, national, xxiv, 179, 359
Interior lines. *See* Line, interior; Communications
International Law, 358
International relations, change in structure of, xvii
  See also Retrospect and Prospect: Studies in International Relations, Naval and Political
Iran. *See* Persia
Iraq. *See* Mesopotamia
Ireland, 194, 197, 296, 299
Irish Sea, 43, 297
Italy, 371
  and geographical position, 34–35, 42, 43, 306

Jamaica, 13, 113, 119, 171, 178, 179, 186, 204, 229, 230, 243, 246
Japan, xvii, 158, 172, 175, 191, 294, 305, 307, 308, 314, 341
  geographical position, 259
  Inland Sea of, 178
  map, 148
  See also Wars, Russo-Japanese
Jomini, Henri, 105, 106, 217, 231, 247, 277
  and ATM, ix, xxxi
  and Dennis Mahan, xiii
  quoted, 23, 108, 127, 262, 278–79

Kamranh Bay, 249
Key, of a military situation, 99, 227, 231
Key West, 37, 106, 107, 112, 138, 143, 173, 183, 244
Kiel Canal, 381
Kitchener, Field Marshal Lord, 162, 164
  quoted, 160
Kure, 178

Latouche Tréville, Admiral, 214
Laughton, Sir John Knox, x, xi, xvii
Law. *See* International Law
Le Havre, 312
Lepanto, 14
Lessons of the War With Spain, quote from, xix, xxi, xxii, xxix
Levant, 14, 35, 40, 48, 114, 178, 308, 334, 335, 371
Life of Nelson, quote from, xxii, xxvii,
Line, interior, xxvi, 21, 235, 266, 318, 367
  of battle, 257, 350
  strategic, 142–76, 229, 244
  See also Communications
Logistics, 97–98, 105, 107, 138, 188, 189, 207, 231, 244, 245, 251, 269, 291, 344–46, 355, 358, 381
  See also Resources
Louisburg, 30, 171, 232
Luce, Stephen B., x, xi, xiii, xiv, xvii

Madagascar, 249
Mahan, A. T.,
  articles and books by. *See* title names
  biographical sketch, xii–xv
  contemporary opinion on, xv–xvi
  contribution to naval thought, ii, ix–xii, xxxi–xxxii
  intellectual world of, xv–xviii
  Naval War College, xiv
  portrait, frontispiece

Mahan, Dennis H., ATM's father, xiii
Mail (Tri-weekly Times), quote from, 160
Major Operations of the Navies in the War of the American Revolution, quote from, xxii
Malta, 30, 35, 111, 112, 115, 157, 158, 163, 164, 165, 166, 171, 178, 179, 187, 190, 198, 204, 209, 212, 214, 215, 218, 229, 230, 253, 314, 316, 327, 334
Manchester Guardian, quoted from, xv-xvi
Maritime region, 181. *See also* Theater of war
Marlborough, Duke of, 233, 234
Marseille, 312
Martinique, 34, 80, 81, 119, 140, 236, 253
Mauritius, 30
Maxims. *See* Doctrine; Principles; Teachers
Mediterranean, 13, 14, 22, 25, 31, 34–36, 39, 45, 64, 102, 103, 114, 118, 147, 157, 162, 163, 165, 166, 171, 175, 179, 183, 185–87, 208, 215, 223, 229, 233–39, 253, 262–64, 268, 288, 293, 297, 300, 304–14, 327, 335, 340, 365, 366, 367, 369, 371, 373
Mêlée, 3–4
Merchant shipping. *See* Shipping, merchant
Mesopotamia, 328, 334, 337–39
Mexico, 372
Mine warfare, 122, 124, 134–35, 137
Minorca, 111, 157, 158, 166, 171, 234, 263, 292
Mission, 354, 356
Mississippi River, 36, 37, 39, 42, 47, 106, 114, 143, 151, 183, 209
Mobile Bay, 151, 174
Mobility, xxii, xxiii, xxv, 23, 105, 123, 167, 231, 282, 290, 295, 301, 334, 344
Mobilization, 104–5
Mommsen, 15; quoted, 17
Mona Passage, 230
Monitors, 129–30
Monroe Doctrine, 174, 175, 190, 286, 359, 360, 372
Moore, General Sir John, 217, 218, 224, 248, 252, 352–53
Moral effects, xxii
Morogues, Bigot de, 10, 83

Napier, General, 218, 248
Naples, 163, 214, 263
Napoleon Bonaparte, 2, 12, 23, 50, 115, 118, 121, 143, 145, 147, 156, 164, 165, 170, 180, 185, 190, 191, 192, 213, 219, 231, 255, 264, 285, 351, 352, 357, 369, 379
  quoted, xxx, 105, 138, 161, 165, 188–89, 197, 209, 218, 221–22, 224, 237, 267–68, 276, 277, 280, 378
Narragansett Bay, 149, 150, 252
National character. *See* Character, national
National interests. *See* Interests, national
National policy. *See* Policy
National Review, ATM article in, 281–318, 319–42
Naval Administration and Warfare, quote from, xxv
Naval Strategy, x
  lectures commented on by Gen. Sherman, 103
  original lectures written 1887, 133
  quotes from xii, xxi, xxiii, xxvi, xxviii, xxx, xxxi; section, 233–35; chapter VI, 97–110;

Naval Strategy (*continued*)
chapter VII, 111–41; chapter VIII, 142–76; chapter IX, 177–218; chapter X, 219–80
Naval strategy. *See* Strategy, naval
Naval strength and superiority. *See* Strength
Naval War College, x, xiv, xv
ATM on, 98, 343–58
ATM's debt to, xiv
war games at, 104
Navarino, 14n.5, 199
Navy, function of, 9, 28–29, 94n, 131, 176, 269
See also Fleet
Navy yard. *See* Dockyards
Nebogatoff, Admiral, 249
Nelson, Horatio, 9, 12, 24–26, 65, 98, 115, 116, 125, 135, 136, 137, 144, 163, 178, 185, 191, 195, 213–16, 223, 224, 237, 250, 258, 263, 265, 267, 302, 314, 344, 350, 357
quoted, 195, 247, 248, 267, 270, 309–10, 313, 348, 377
See also Life of Nelson
Nelson, Jean Ware, xviiin.13
Netherlands, xxx, 27, 65, 285, 370
and character of government, 72–74
and geographical position, 31–37, 114, 163
and national character, 55–61
and physical conformation, 38–45, 153
New Orleans, 173, 209
New York, 30, 39, 114, 136, 144, 147, 149, 174, 183, 225, 252, 271, 272
New Zealand, 158, 160, 367
Nice, 236, 237
Nile, battle of, 11–12, 116, 159, 161, 164, 191, 195, 211, 213–17, 233, 235, 236, 239
map, 240–41
See also Egypt
Norfolk, 149, 174, 183, 252, 271, 272
North American Review, articles by ATM from, 343–58, 368–81
North Sea, 34, 108, 109, 161, 183, 189, 296, 297, 298, 299, 316, 340, 365

Objective and objectives in war, xxiv–xxv, xxvii, xxx, 9, 154, 176, 180, 182, 190, 196, 219, 288, 298
Occupation, of a country, 24
Offensive, xviii, xx, xxi, xxii, xxv, xxvi, 111, 113, 126, 128, 131, 146, 157, 164, 167, 180, 181, 182, 196, 219, 255, 259, 262, 274, 288, 289, 290, 294, 295, 297, 298, 304
and concentration, 265, 311
offensive-defensive, 121, 126, 219, 224
offensive strength, 131–40
Open Door Policy, 174
Operations, combined, 181, 192, 219
Operations of war, 8, 180, 219–80
Orders, formulation of, 349–50, 354
Orkney Islands, 296
Oxford University, xv

Pacific Ocean, xx, xxviii, 118, 175, 185, 188, 198, 287, 304, 330, 372, 375–77
Panama Canal, xx, 29, 35, 36, 45, 57, 94, 106, 107, 114, 118, 119, 138, 175, 179, 186, 188, 190, 192, 230, 244, 245, 259, 305, 334

"Panama Canal and the Distribution of the Fleet," 368–81
Pax Britannica, xvi
Peace, and naval strategy needed during, xxvi, 294, 355, 358, 362, 371–72
Pellew, Admiral Sir Edward, 49, 303, 310
Pensacola, 37, 106, 112, 138, 151, 173
People, character of. *See* Character, national
Persia, 308
Persian Gulf, 307
  article "Persian Gulf and International Relations," 319–42
Philippines, 172, 377
Plymouth, 138, 171, 183, 229
Policy, national, xx, xxiii, 170, 322–23, 358
  and warfare, xxxi
Political science, study of, ix
Pondicherry, 13
Population, number of, 31, 47–53
Port Arthur, 120–22, 127, 134, 136, 139, 149, 152, 155, 158, 166, 168, 169, 170, 178, 191, 205, 225, 230, 243, 248, 253, 270, 271, 325, 332, 353, 380
Port Mahon, 229
Porto Rico, xx, 112, 172, 243, 244, 259
Port Royal, 138, 173, 244
Ports. *See* Bases
Portsmouth, 138, 171, 183, 229
Position, xx, xxv, 23, 111–19, 173, 193, 231, 305, 313, 315, 357, 369
  central, 21, 266, 270, 298, 315, 316, 318, 371, 380
  decisive, 250
  geographical, 31–37, 140, 162, 282
  the sea as, xxvi, 155, 235, 268
  and strategic points, 105–10, 140, 153, 179, 189–90, 211, 245, 255, 293
Precedents, different from principles, 7, 351
Preparedness, xxvi
Principles, military and naval, x, xii, xxxi, 2, 7, 10–11, 14, 97–110, 211, 216, 227, 273, 275, 278–80, 351, 368
Privateers, 187. *See also Guerre de course*
Problem of Asia, quote from, xxviii
Problem of Asia and its Effect Upon International Policies, quote from, xix
Production, 30
Property at sea, private, xxv, 284
Puerto Rico. *See* Porto Rico
Puget Sound Navy Yard, xv, 136
Pursuit, 246, 266, 267

Quebec, 156, 171, 232

Raiding, 228, 244, 301
  See also Guerre de course; Shipping, merchant
Railroads, 324–25, 332–33, 335, 355
Rams, 133–34, 203
Red Sea, 314, 323
Reserve force, as staying power, 48–52
Resources, 111, 140–41, 139, 173, 282, 315
  See also Logistics
Retaliation, xxvi
Retreat, 246, 249, 250, 252, 265, 266, 272, 380
Retrospect and Prospect, quote from, xx, xxv; chapter from, 281–318, 319–42
Rhine River, 150, 163, 234

Richelieu, quoted, 63–64
Risk, xxvii, 153, 357. *See also* Chance
Rochefort, 161, 187, 220, 228
Rodgers, Commodore John, 225, 272
Rodney, Admiral Lord, 7, 119, 157, 242, 243, 246, 317, 348, 349, 368
  quoted, 112, 115, 349
Rome, 15–22, 53, 363
Rosyth, 104, 171, 183
Rozhestvensky, Admiral, 98, 142, 143, 145, 170, 191, 204, 248, 249, 253, 380
Russia, xxviii, 14, 67, 101, 158, 178, 181, 191, 265, 287, 305–8, 311, 314, 328, 329, 332, 336, 342, 369
  See also Wars, Russo-Japanese

St. Helena, 30
St. Lucia, 140, 179, 196, 230, 236, 243, 244
St. Privat, battle of, 348
St. Vincent, Admiral Lord, 9, 11, 135, 302
  quoted, 137, 163
Samana Bay, 230, 244
San Francisco, 149
Santiago, 121, 145, 146, 151, 178, 186, 205, 225, 231, 245, 257, 378
Schiemann, Theodore, quoted, 366–67
Schurman, D. M., xn.2
Scipio, 16
Scotland, 297, 299
Sea, as a bridge, 235
  as a medium of circulation, 118, 293, 323–24
  mastery of, 232
Sea power, xxiii, xxvii–xxviii, 14, 17, 18, 22, 23, 95
  based on merchant shipping, 53, 76, 118
  elements of, 27–96
  noiseless pressure of, xxviii, 68
Sea Power in its Relation to the War of 1812, quote from, xxvi, xxvii, xxx
Sebastopol, 158, 168, 181
Shakespeare, William, quoted, 363
Sherman, General W. T., 156, 209
  reviewed ATM's lectures, 103
Ship names, USS *Chicago,* xv; USS *Constitution,* 274; HMS *Guerrière,* 274; USS *Macedonian,* xiii; USS *Massachusetts,* 245; USS *Wachusett,* xiii; USS *Wasp,* xiii
Shipping, merchant, xviii, xxiii, xxv, xxvii, xxviii, xxx, 1, 9, 27–31, 40–41, 53, 76, 107, 117, 161, 272, 284–88, 293, 296, 304, 312, 315, 365
Ships
  supply, xxix, 188, 345
  tonnage of warships, xxi, xxxi
  types of warships, 2–7, 23, 126, 129
  warships, xxxi
  See also Steamships; Galleys, Monitors, Torpedo-vessels
Sicily, 163, 263, 267, 270, 306, map, 206
Singapore, 333
Some Neglected Aspects of War, quote from, xxii
South Africa, 158. *See also* War, Boer
South America, 287
Spain, 369, 378
  and character of government, 66–68
  map, 226
  and national character, 53–55
  and physical conformation, 43–44

See Wars, and *Lessons of the War with Spain*
Speed, xxi, xxii, xxix, xxxi, 188, 245, 268, 269, 301
See also Mobility
Statesmen and Statesmanship, xxiv, 102, 281–83, 322, 362
sea power of interest to, 24, 375–76
Steamships, 2–7, 23, 97, 152, 301
Strategic points, xxiii. *See* Position
Stratego-tactical, 357
Strategy, elements of, xix, 9, 23, 232, 282, 344, 346, 358
naval, xxiv, xxvi, xxix, 23, 95, 100–101, 184, 305
naval, in peacetime, 102–3
Strength
military, 111–40, 173, 282
naval, xxiii, 119–41, 177, 184, 255
and naval inferiority, 196, 246, 247, 259, 293
and naval superiority, 186, 190, 246, 262, 275, 293, 295, 297, 299, 311, 368, 371
Study. *See* History, study of
"Submarine and its Enemies," quoted from, xxx
Submarines, xxx
Suez Canal, 102, 114, 118, 190, 230, 288, 300, 314, 316, 323, 332, 334, 335, 371
Suffren, Admiral, 239, 242, 251
Superiority. *See* Strength
Supplies. *See* Logistics

Tactics, xxiv, 2–7, 9–12, 83, 95, 133–35, 136, 196, 203, 210, 269, 276, 283, 344, 346, 357, 358
grand tactics, 213, 269
Teachers, xxiii, 279
Territory, extent of, 31, 45–47
Theater of war, 8, 99n.1, 110, 128, 153, 177, 231, 250, 251, 292
Togo, Admiral, 116, 143, 147, 248, 249, 252, 259, 270
quoted, 139, 191
Tonnage, of ships, xxi
Torpedo, xxx, 153
Torpedo-vessels, 126, 152, 183, 301
Toulon, 147, 161, 178, 187, 192, 209, 211, 212, 214, 220, 228, 233, 234, 235, 236, 237, 270, 305, 310, 312, 333
Tourville, Admiral, 247
Trade. *See* Shipping, merchant
Trafalgar, 9, 24–26, 50, 86, 87, 137, 159, 346, 370
Tsushima, battle of, 116, 249, 252, 257
See also Togo; Wars, Russo-Japanese
Tunis, 102
Turkey, 14, 40, 51, 101, 171, 214, 326, 327, 335
"Two Maritime Expeditions," article by ATM reprinted, 198–218

Uncertainty. *See* Chance
United Service Magazine, article by ATM reprinted from, 198–218
United States of America
and Asiatic immigration, 174–75
and geographical position, 32, 36–37
and national character, 61–62
and physical conformation, 41–43, 45
and population, 52–53
and the sea, xxvi, xxviii, 29, 89–94, 123, 229, 259

Venezuela, 175
Victoria, Queen of England, xv
Victory, xx, xxix, 23, 50, 192, 193, 238, 251, 279, 289
Vistula, 13
Vladivostock, 178, 248, 249, 252, 257, 270

Warfare
- art of, 276–80, 370
- and changed circumstances of modern, 90, 95–96, 122, 172, 272, 301–3, 362–63, 375
- civilized, xxv
- land, in relation to naval, 94n, 100, 182, 184, 260, 291
- learning from experience of, 275–80. *See also* History
- nature of, xxvi, xxx–xxxi
- as a political act, xxxi, 283

War games, 356–57
Wars
- American Civil (Secession), xv, 39, 46, 47, 94n, 103, 108, 138, 145, 169, 173, 174, 185, 244. *See also Gulf and Inland Waters; Admiral Farragut*
- American Revolution, xxix, 13, 26, 38, 70, 83–86, 112, 159, 166, 168, 176, 180, 181, 186, 194, 232, 239, 243, 251, 320. *See also Major Operations of the Navies . . .*
- Anglo-Dutch, 153, 285, 296, 370; First, 40–41, 365; Second, 38; Third, 73; Fourth. *See* American Revolution
- Austrian Succession, 68–69
- Boer, 170, 312, 334
- Crimean War, 158, 185
- Franco-Prussian, 193, 348
- French Revolution and Empire, xxviii, 28, 39, 49, 71, 86, 87, 112, 116, 136, 137, 138, 144, 145, 159, 161, 162–65, 170, 177–78, 179, 180, 185, 191, 192, 194–98, 200, 207–18, 220–23, 228, 233, 243, 247, 248, 252–56, 267–68, 320, 345–46, 348, 352–53, 357. *See also Influence of Sea Power Upon the . . .*
- Mexican, 185
- Nine Years, 73, 194, 195, 247, 365
- Peloponnesian, 199–207
- Punic, 14–23
- Russo-Japanese, 98, 116, 120–22, 124, 126–28, 134, 136, 137, 138, 139, 142–43, 152, 153, 158, 166, 169, 170, 178, 191, 198, 204, 230, 243, 248, 249, 252, 257, 272, 275, 353, 355–56, 370, 372–73
- Seven Years, 69, 70, 81, 82, 172, 191, 192, 194, 224, 232, 243, 251
- Spanish-American, xv, 122, 128–29, 145, 146, 205, 231, 245, 257–58, 273, 378. *See also Lessons of the War with Spain*
- Spanish Succession, xxviii, 69, 233–35
- Thirty Years, 38
- War of 1812, 38, 143–44, 203, 224, 225, 235, 272
- World War I, xv, 369–72, 381. *See also Sea Power in its Relation to the War of 1812*

Washington, George, quoted, 319, 320, 322, 331, 342
Wealth of states, xxvii, xxx, 363, 368
Weather, 23, 272
Wellington, Duke of, 220, 221, 222, 235, 248, 378

West Indies, 64, 112, 113, 115, 117, 118, 119, 176, 185, 194, 195, 222, 239, 253, 274, 345, 346
Wilhelm II, Kaiser, xv, 366, 376
Wood, General Leonard, 123

Yorktown, battle of, 242

CLASSICS OF SEA POWER

John B. Hattendorf, Wayne P. Hughes, Jr.,
Series Editors

Sir Julian Corbett, *Some Principles of Maritime Strategy.* Introduction by Eric J. Grove

Bradley Fiske, *The Navy as a Fighting Machine.* Introduction by Wayne P. Hughes, Jr.

Wolfgang Wegener, *The Naval Strategy of the World War.* Translation and introduction by Holger H. Herwig

J. C. Wylie, *A General Theory of Control.* Introduction by John B. Hattendorf

Vice Admiral P. H. Colomb, RN, *Naval Warfare: Its Ruling Principles and Practice Historically Treated,* Two Volumes. Introduction by Barry M. Gough

S. O. Makarov, *Discussions of Questions in Naval Tactics.* Translated by J. B. Bernadou. Introduction by Robert B. Bathurst

ABOUT THE EDITOR

John B. Hattendorf has been the Ernest J. King Professor of Maritime History at the Naval War College since 1983. He is the senior editor of the Classics of Sea Power series.

After receiving his bachelor's degree in history from Kenyon College in 1964, Hattendorf served at sea in both Atlantic and Pacific Fleet destroyers, seeing combat action in Vietnam. Later he served ashore at the Naval Historical Center in Washington and on the staff of the Naval War College. After leaving active duty, he obtained a master's degree in history at Brown University and went on to obtain his doctorate at the University of Oxford in England.

The editor of a number of volumes of naval documents as well as articles on naval history and strategy, among his recent publications are *England in the War of the Spanish Succession;* the introduction to J. C. Wylie's *Military Strategy* in the Classics of Sea Power series; *A Bibliography of the Writings of Alfred Thayer Mahan* (compiled with Lynn C. Hattendorf, 1986/1990); *Maritime Strategy and the Balance of Power: Britain and America in the 20th Century* (co-edited with Robert S. Jordan, 1989); *The Limitations of Military Power: Essays Presented to Professor Norman Gibbs* (co-edited with Malcolm H. Murfett, 1990); and *The Influence of History on Mahan: Proceedings of the Mahan Centennial Conference (1991).*

The Naval Institute Press is the book-publishing arm of the U.S. Naval Institute, a private, nonprofit, membership society for sea service professionals and others who share an interest in naval and maritime affairs. Established in 1873 at the U.S. Naval Academy in Annapolis, Maryland, where its offices remain today, the Naval Institute has members worldwide.

Members of the Naval Institute support the education programs of the society and receive the influential monthly magazine *Proceedings* or the colorful bimonthly magazine *Naval History* and discounts on fine nautical prints and on ship and aircraft photos. They also have access to the transcripts of the Institute's Oral History Program and get discounted admission to any of the Institute-sponsored seminars offered around the country.

The Naval Institute's book-publishing program, begun in 1898 with basic guides to naval practices, has broadened its scope to include books of more general interest. Now the Naval Institute Press publishes about seventy titles each year, ranging from how-to books on boating and navigation to battle histories, biographies, ship and aircraft guides, and novels. Institute members receive significant discounts on the Press's more than eight hundred books in print.

Full-time students are eligible for special half-price membership rates. Life memberships are also available.

For a free catalog describing Naval Institute Press books currently available, and for further information about joining the U.S. Naval Institute, please write to:

Member Services
U.S. Naval Institute
291 Wood Road
Annapolis, MD 21402-5034
Telephone: (800) 233-8764
Fax: (410) 571-1703
Web address: www.usni.org